绿色建筑产品认证与评价
——国内外典型政策和案例解读

佟晓超　编著

中国建筑工业出版社

图书在版编目（CIP）数据

绿色建筑产品认证与评价——国内外典型政策和案例解读/佟晓超编著．—北京：中国建筑工业出版社，2017.4
ISBN 978-7-112-20605-6

Ⅰ.①绿… Ⅱ.①佟… Ⅲ.①生态建筑-产品质量认证-研究②生态建筑-产品质量-评估-研究 Ⅳ.①TU-023

中国版本图书馆CIP数据核字（2017）第061347号

责任编辑：费海玲　张幼平
责任校对：李美娜　焦　乐

绿色建筑产品认证与评价——国内外典型政策和案例解读
佟晓超　编著

*

中国建筑工业出版社出版、发行（北京海淀三里河路9号）
各地新华书店、建筑书店经销
北京科地亚盟排版公司制版
北京市密东印刷有限公司印刷

*

开本：787×1092毫米　1/16　印张：13¼　字数：294千字
2017年4月第一版　2017年4月第一次印刷
定价：**38.00**元
ISBN 978-7-112-20605-6
（30271）

版权所有　翻印必究
如有印装质量问题，可寄本社退换
（邮政编码　100037）

前　言

我国近些年来，绿色建筑发展迅猛，绿色建筑产品的市场需求凸显。鉴于绝大多数情况下，最终用户无法影响建造阶段对于建筑产品的选用，政府如何在简政放权的模式下，作为"掌舵人"做好准公共产品的管理工作，通过政策、法律和行政性规章的制定，充分调动和利用社会第三方机构资源，运用标准化和云技术，系统解决绿色建筑产品性能评价技术、绿色建筑产品数据库与绿色建筑设计数据库的耦合技术；通过综合运用新公共管理和政策执行模式理论等，基于委托—代理模型，建立信息不对称情况下风险分担和监管机制，实现对"划桨人"的风险可控，从而探索绿色产业社会效益和经济效益的综合量化评价方式。本书介绍了产品认证、绿色产品和政策制定有关概念和案例，供感兴趣的专家学者和利益相关方品读。为了读者更好地了解国外的研究水平，附录中的外文文献以原貌呈现。因本人水平有限，在撰写中会有疏漏和错误之处，敬请读者指正。

在本书撰写调研过程中，得到了江苏省住房和城乡建设厅科技发展中心、内蒙古自治区建设科技推广中心、中国建筑业协会工程建设质量监督与检测分会、德国建筑技术研究院（DIBt）等单位的大力配合和支持，在此一并致谢！

佟晓超

中国建筑科学研究院

2017.1

目　录

前言
第一章　概述 ··· 1
第二章　产品认证发展 ·· 2
 第一节　产品认证的起源 ·· 2
 第二节　认证概念 ·· 4
 第三节　我国认证的发展 ·· 11
 第四节　我国产品认证与政策、法律和标准的关系 ············ 16
 第五节　我国绿色建筑产品市场现状 ································· 18
第三章　我国建筑产品管理 ··· 22
 第一节　国家政策 ·· 22
 第二节　产业政策 ·· 24
第四章　国外绿色建筑产品 ··· 31
 第一节　德国 ··· 31
 第二节　英国 ··· 37
 第三节　美国 ··· 38
 第四节　新加坡 ·· 46
 第五节　碳足迹 ·· 47
 第六节　环境产品声明 EPD ··· 52
第五章　我国绿色建筑产品评价 ·· 61
 第一节　产品、部品和建材的概念 ··································· 61
 第二节　绿色产品的概念和特征 ······································· 62
 第三节　绿色建筑产品评价体系 ······································· 64
 第四节　构建我国绿色建筑产品评价体系建议 ·················· 66
第六章　认证制度与政策、法律和法规 ·································· 71
 第一节　社会管理理论 ··· 71
 第二节　新公共管理理论 ·· 73
 第三节　政策执行过程模式 ·· 74
 第四节　我国绿色建筑产品认证的政策环境 ······················ 75
第七章　认证制度与标准体系 ·· 80
 第一节　现有标准体系框架简介 ······································· 80
 第二节　绿色建筑产品标准体系探讨 ································· 81

第三节　绿色建筑产品与绿色建筑 …………………………………… 83
第八章　认证制度与政府管理 ……………………………………………… 84
　　第一节　风险概念 …………………………………………………… 84
　　第二节　委托—代理理论 …………………………………………… 85
　　第三节　理论解决方案 ……………………………………………… 87
　　第四节　构建绿色产品认证制度探讨 ……………………………… 88
附录　建筑用基础材料碳排放数据库示例 ………………………………… 90
　　　混凝土 EPD 评价采用的 PCR 示例 ……………………………… 111
　　　GreenFormat 第一版和第二版（草案）示例 …………………… 189

第一章 概 述

——吾生亦有涯，而知亦无涯

随着我国经济的发展，参与乃至主导国际化工作的增多，绿色建筑产品认证和评价作为一个很专业的领域在国外和国内都有来自不同领域的专业人员从事相关的研究工作，并推动其发展。

就绿色建筑产品认证政策研究而言，涉及合格评定、社会管理和政策执行、绿色标准和评价体系等多学科的交叉研究。面对我国众多利益相关方对产品认证的认知、行业政策的导向、市场供需的规范、国外发展的趋势、产品与建筑的结合、标准体系的构建和政府采信的开展等诸多疑惑，本书分八章论述有关概念并通过示例予以说明。附录中有3个国外文献，分别为英国建筑用基础材料碳排放数据库、欧洲混凝土进行环境产品声明（EPD）评价时依据的产品类别规则（PCR）以及美国绿色建筑产品信息标准化规范GreenFormat。

本书的主要内容框架和各章关系如图1-1所示。

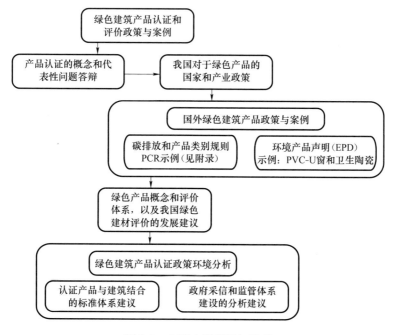

图1-1 主要内容框架与关系

第二章 产品认证发展

——千里之行，始于足下

本章主要阐述了如下内容：
- 产品认证起源：介绍了从18世纪德国建筑用砖的质量问题到20世纪英国BS认证出现的背景，以欧盟、WTO和ISO为例，着重讲述区域化、国际化和标准化组织对产品认证的需求和推动作用。
- 认证有关概念：主要讲述了合格评定、认证认可、检验检测的相关概念；相互之间的关系与差异性。重点回答了利益相关方对产品认证误区的7个代表性问题。
- 我国认证的发展历程：介绍了我国认证领域发展的情况和有关政策。
- 认证与政策、法律和标准的关系：以我国近些年"五年规划"为主线，阐述了产品认证在我国经济中的地位和作用，上位法的依据以及和标准的关系，介绍了NQI和认证行业标准。
- 绿色建筑产品市场现状：以《生态文明体制改革总体方案》为指导思想，以建筑产品自身特征为依据，分析了我国绿色建筑产品市场的需求和发展要求。

第一节 产品认证的起源

自古以来，衣食住行都与民生息息相关，是国家稳定和发展的重要内容。在古代，与"衣"、"食"、"行"相比，"住"从资产的投入和所需的建造技术要求都是最高的。正因为投入大，技术要求高，一旦因产品质量出现问题引发工程事故，会造成人员伤亡和财产的损失。建筑工程通常建造周期长，隐蔽工程多，为了解决建筑工程质量问题，控制工程中使用的建筑产品质量符合要求是重要环节之一。

在1738年，建筑用砖成为德国工程中使用的大宗建材，但在当时，由于生产工艺和配套的生产设备设施相对简陋，导致生产出的成品砖的质量很不稳定，工程现场有大量的残次品，对建筑质量造成了严重影响。鉴于此，当时德国地方政府规定，建筑用砖在投入工程使用以前，必须出示政府出具的质量检测报告。这种形式的出现是合格评定制度的雏形。

19世纪末，随着市场经济的发展和科学技术的发展，新产品层出不穷，并且越来越复杂，普通消费者在购买产品时已经不可能自行鉴别产品的质量。政府为保证社会秩序的正常运行，对于涉及人身安全的产品开始实行监管。因此，在当时社会情况下，

第一节 产品认证的起源

政府是自己承担了对部分与人身安全直接相关的产品质量的监管职能。

在20世纪上半叶，不少国家的政府承担了产品质量的监管职能，为了实施监管职能，尝试了不同的形式。1903年，英国创立了世界上第一个符合性认证标志，即使用BS字母组成的"风筝标志"，标志施加在钢轨上，表明钢轨符合质量标准，该标志以英国国家标准为检验依据。1907年德国萨克森地区建立了统一的建筑材料用检测基本规则。1937年底德国建立了建筑产品和施工工艺监管认定体系。在20世纪初，欧洲许多国家纷纷效仿英德，建立起以本国标准为依据的质量管理制度。当时各国对于建筑产品所采用的认证制度还比较简单，通常是一种产品认证制度型式认定制度。可以看出，在欧洲各国的尝试中，关注重点主要是建立标准和在本国范围内统一技术标准，并据此对产品开展符合性认证评价。

第二次世界大战后，世界经济格局发生了重大变化，国际贸易迅速增长，产品在各国间的流通频次和数量急剧增加，为了协调商定双边和多边贸易中各国自身的安全、经济、健康等诸多利益关系，一些区域性和国际性的组织陆续成立。为了解决各国贸易间存在的诸多问题，这些区域性和国际性组织推动了产品认证制度的发展。最具代表性的区域性组织就是欧洲联盟（简称EU或欧盟），与产品认证密切相关的国际性组织有世界贸易组织（简称WTO）和国际标准化组织（简称ISO）。

欧盟的雏形可以追溯到1950年，即欧洲煤钢共同体，当时的标志性事件就是"舒曼计划"的诞生。欧洲煤钢共同体成立7年后，在1957年3月25日，法国、意大利、联邦德国、荷兰、比利时、卢森堡等6国外长在罗马签订了建立欧洲经济共同体与欧洲原子能共同体的两个条约，即《罗马条约》，该条约于1958年1月1日生效。《罗马条约》生效7年后，1965年4月8日，联邦德国、法国、意大利、荷兰、比利时、卢森堡签订了《布鲁塞尔条约》，决定将欧洲煤钢共同体、欧洲原子能共同体和欧洲经济共同体统一起来，统称欧洲共同体，《布鲁塞尔条约》于1967年7月1日生效。1991年12月11日，欧洲共同体举行首脑会议，通过了建立欧洲经济货币联盟和欧洲政治联盟的《欧洲联盟条约》。1992年2月7日，各国外长正式签署《欧洲联盟条约》。经欧洲共同体各成员国批准，《欧洲联盟条约》于1993年11月1日正式生效，欧洲共同体开始向欧盟过渡。

世界贸易组织可以追溯到1946年，当时由美、英等19个国家组成的联合国贸易与就业会议筹备委员会，起草了《联合国国际贸易组织宪章》。1947年10月30日在日内瓦筹备委员会的成员国共同签订了《关税及贸易总协定》（简称GATT），该议定书于1948年1月1日生效。关税及贸易总协定原计划只是在国际贸易组织成立前的一个过渡性步骤，但是，各国对外经济政策方面的分歧和其他法律方面的困难，使得关税及贸易总协定的有效期一再延长，为适应不断变化的情况，各国对关税及贸易总协定进行了多次修订。

直到1994年4月15日，在摩洛哥的马拉喀什市举行的关税及贸易总协定乌拉圭回合部长会议决定成立更具全球性的世界贸易组织（简称WTO），以取代成立于1947年的关税及贸易总协定。世界贸易组织（WTO）与关税及贸易总协定（GATT）是不

第二章　产品认证发展

同的，区别如下：

1) GATT 是临时性的。GATT 从未得到成员国立法机构的批准，其中也没有建立组织的条款。WTO 及其协议是长期的。作为一个国际组织，WTO 具有良好的法律基础，因其成员已经批准 WTO 协议，而且协议规定了 WTO 应如何运作。

2) WTO 拥有"成员"，GATT 拥有"缔约国"，也就是说 GATT 只是一个法律文本。

3) GATT 仅处理货物贸易，WTO 还涉及服务贸易和知识产权内容。

4) WTO 争端解决机制与原 GATT 体制相比，速度更快，作出的裁决不会受到阻挠。

受关税及贸易总协定的影响，1970 年，国际标准化组织（ISO）成立认证委员会，随着合格评定概念的发展，认证认可制度的不断完善，ISO 认证委员会于 1985 年改名为合格评定委员会（简称 CASCO），CASCO 开始从技术角度协调各国认证制度的内容，促进各国认证机构结果的相互认可，以消除各国由于标准、检测、认证过程中存在的差异所带来的贸易困难，并进一步制定出国际性的认证制度。1993 年结束的"乌拉圭回合"谈判，将质量认证扩展为"合格评定程序"。1994 年该委员会又更名为合格评定发展委员会。合格评定研究和管理领域目前主要包括产品认证、管理体系认证、认证机构认可、检测和检查机构认可等内容。

以欧盟、世界贸易组织和合格评定发展委员会及其前身的建立为标志，世界经济呈现出国际和区域化的发展趋势。伴随进出口贸易的增长，单一国家内部执行的产品认证制度的局限性逐步暴露出来，欧盟首先出现以区域性标准为依据的双边和多边的产品认证制度，其中最有影响力的就是欧洲强制性产品认证制度，即 CE 认证。CE 认证的技术依据是欧洲标准委员会和欧洲技术评估组织发布的，各成员国都接受的技术标准和技术规范。这种区域性产品认证制度的建立，克服了欧盟各成员国之间原来标准不统一和管理技术上的差异造成的障碍，简化了欧盟区内的贸易手续，保护了各成员国的利益。但是对非欧盟成员国的技术壁垒依然存在。这种背景促进了产品认证国际互认的需求和认可的发展。

第二节　认证概念

产品认证属于合格评定的一种形式。什么是合格评定？根据国际标准化组织合格评定委员会（CASCO）的定义，合格评定是指与产品、过程、体系、人员或机构有关的规定要求得到满足的证实，即任何直接或间接确定技术法规或标准中相关要求被满足的一整套的程序和管理制度。换言之，合格评定的定义包含如下三方面内容：

第一，合格评定的对象可以是产品、过程、体系、人员或机构。

第二，合格评定的作用是证实有关规定要求是否得到满足，这里提到的有关规定可以是技术法规或标准。

第三,合格评定的构成是一整套的程序和管理制度。

合格评定包括认证和认可两大类活动。认可是证明合格评定机构具备实施特定合格评定工作能力的第三方证明。认可的概念包括如下三方面内容:

第一,认可的对象是合格评定机构。合格评定机构是指从事合格评定服务的机构,这里提到的合格评定服务包括认证、检查、检测等活动。

第二,认可评价的是从事合格评定服务机构的能力。也就是说,认可评价的是认证机构、检查机构和检测机构的能力。

第三,认可是第三方组织出具的证明。这里提到的第三方组织是一个特定的从事认可评价活动的第三方组织。在国际上,这类从事认可活动的第三方组织不一定是唯一的,但一定是经过所在国家主管部门依据法律法规进行授权的。我国合格评定行业的主管部门是中国国家认证认可监督管理委员会(简称国家认监委,CNCA)。目前,我国只有唯一一家经国家认监委授权从事认可的机构,即中国合格评定国家认可委员会(简称认可委,CNAS)。

在合格评定中除了认可活动以外,真正与社会生活以及生产直接相关且量大面广的是认证活动。认证是指与产品、过程、服务、体系或人员有关的第三方证明。由上述定义可以看出,认证的对象不同于认可的对象。认证的对象是产品、过程、服务、体系或人员。根据认证对象的不同可以将认证分为几种,见图2-1。

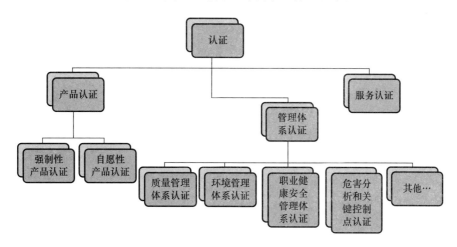

图2-1 认证类型划分

认证同样是第三方机构出具的证明。但这里的第三方机构是从事认证活动的认证机构。现将上述合格评定、认可、认证等一系列概念的相互关系描述如图2-2。

产品认证是由第三方机构证明产品及其实现过程具有某种特定的功能和特性,并符合标准和其他规范性文件的规定要求的合格评定活动。产品认证的概念包括:第一,产品认证是由第三方机构实施的;第二,产品认证的对象是产品(或过程);第三,产品认证的作用是符合性证明,即产品功能与标准和其他规范性文件规定的要求相符。

产品认证通过与特定产品相关的,适用相同规定要求、规则和程序的认证方案来实现。比如有些产品的认证方案可包括初始检测和对供方质量管理体系的评审,以及

第二章 产品认证发展

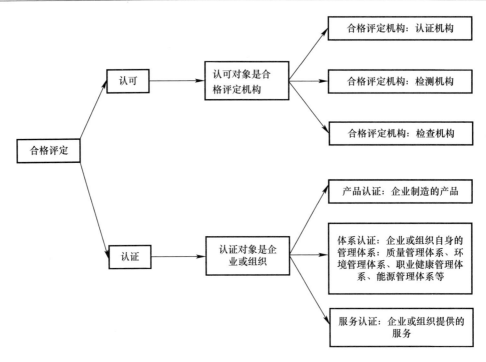

图 2-2 合格评定与认证认可的关系

后续的监督几部分内容。对于后续监督，要考虑对质量管理体系的评审以及从生产现场和市场抽样检测或检查。另外，也有些认证方案基于初始检测和监督检测，还有一些认证方案仅包括型式试验。认证方案是产品认证制度在具体产品上的体现，认证方案的多样化是依据产品认证制度而定的。根据 2004 年 ISO/IEC 发布的指南 67，产品认证制度的组成要素包括 ISO/IEC17000 功能法中提到的 5 个部分，即选取（selection）、确定（determination）、审查（review）、证明（attestation）、监督（surveillance）。常见的产品认证制度见表 2-1。

产品认证制度要素　　　　　　　　　　　　　　表 2-1

产品认证制度的要素	产品认证制度					
	1	2	3	4	5	6
1. 选取	√	√	√	√	√	
2. 确定： a）检测 b）检查 c）设计评价 d）服务评定	√	√	√	√	√	√
3. 审查	√	√	√	√	√	√
4. 认证决定： 批准、保持、扩大、暂停、撤销认证	√	√	√	√	√	√
5. 证明： 批准、保持、扩大、暂停、撤销使用证书或标志的权利		√	√	√	√	√

第二节 认证概念

续表

产品认证制度的要素	产品认证制度					
	1	2	3	4	5	6
6. 监督： a) 从公开市场抽样检测或检查 b) 从工厂抽样检测或检查 c) 结合随机检测或检查的质量体系审核 d) 对生产过程或服务评定		√	√ √	√ √ √	√ √ √ √	 √ √

产品认证制度，是指由依法取得产品认证资格的认证机构，依据有关的产品标准和/或技术要求，按照规定的程序，对申请认证的产品进行工厂检查和产品检验等评价工作，对符合条件和要求的产品，在经过认证决定后，通过颁发认证证书和认证标志以证明该产品符合相应标准要求的制度。

根据上面对于产品认证制度的描述，如下要点需要说明：

第一，认证机构。从事产品认证的机构是经过国家认证认可监督管理委员会批准授权后才能在其批准专业范围内从事产品认证工作。

第二，认证依据。对于产品认证而言，认证的核心依据是认证方案，在认证方案中规定了认证模式、认证单元划分、产品性能要求、检测依据、工厂检查内容要求、持续监督要求、认证证书和认证标志规定等内容。

第三，评价工作。评价工作是产品认证的核心技术工作。根据认证方案中规定的认证模式，评价工作可以包括对于产品设计的评审、工厂检查、见证试验、抽样后送第三方试验室进行规定的检测、后续监督等内容。

第四，认证证书和标志。认证证书是申请方通过产品认证评价后获得的书面证明。认证标志可以施加在获证产品或其宣传材料上表明该产品通过了认证。

我经常遇到来自社会各方关于产品认证咨询的问题，归纳起来主要是以下几个：

第一，产品认证和质量管理体系认证有什么不同？

第二，产品认证和检测有什么不同？

第三，产品认证是否等同于质量管理体系认证和检测的集合？

第四，为什么不同认证机构做产品认证的要求不一样？

第五，如何看懂认证证书？

第六，如果产品认证不是强制性的，为什么有采信价值？

第七，产品质量是生产出来的，不是认证出来的，认证有什么用？

上面7个问题集中反映了在住建领域产品认证的概念和基础知识普及度还远远不够。产品认证所承担的社会功能就是提高透明度，传递信任。而今，在众多情况下，普及度尚且不足，何谈传递信任！

第一，对于产品认证与质量管理体系认证的差异主要有如下几点：

1. 认证对象不同：产品认证的对象是企业制造的某一有形产品，如槽式预埋件、混凝土用建筑锚栓、钢筋机械连接接头、塑料管材管件、密封胶等；质量管理体系认证的对象是企业的管理体系，与产品质量无直接关系。

2. 认证依据不同：依据国家认监委的规定，产品认证的技术依据包括产品标准（国家标准、行业标准、国际和国外标准、地方标准、社团标准）或认证技术规范（须在国家认监委备案）和产品认证方案（须在国家认监委备案）；体系认证依据是有关管理体系标准，如《质量管理体系 要求》GB/T 19001，《环境管理体系 要求及使用指南》GB/T 24001，《职业健康安全管理体系 规范》GB/T 28001等。如认证机构可以根据《空气-空气能量回收装置》GB/T 21087—2007做热回收装置的产品认证，也可以根据《职业健康安全管理体系 规范》GB/T 28001做该企业在职业健康安全管理方面的体系认证。但这显然是两件截然不同的评价工作。

3. 认证结论不同：产品认证证明产品的质量或某些性能满足有关产品标准和/或技术规范规定的要求；质量管理体系认证证明企业质量管理体系是否满足《质量管理体系 要求》GB/T 19001标准规定的要求。可以理解为产品认证机构为已获得认证的产品质量进行了担保，而质量管理体系认证并没有直接对产品质量进行担保。

第二，产品认证与产品检测不同，其主要区别如下：

1. 申请的主体不同：产品认证的申请方是制造商或是与制造商有法律关系的授权方；产品检测申请方与制造商可以没有任何关系。比如，对于产品而言，产品的制造商拥有与产品相关的知识产权。产品的生产厂可以是制造商本身，也可能是代工厂。产品的销售商可以是制造商本身，也可能是经销商、代理商或零售商。产品认证的申请方若非制造商时，需要提供制造商委托实施认证申请的证明。对于产品检测而言，委托检测的单位可以是制造商，也可以是生产厂（与制造商不同时），还可以是施工单位、监理单位或其他利益相关方。

2. 评价对象范围不同：产品认证的对象是所有申请认证的产品；产品检测的对象仅为送样产品，抽样产品可以包括所抽样品所属的批次。因此，在检测报告（无论是抽样检测还是委托检测）都会注上仅对来样负责的说明，也就是说，检测报告不能说明该产品生产企业生产的同一品牌同一型号的产品与检测样品的检测结果和结论相同。而在产品认证证书上不会出现这样的表述。

3. 评定依据不同：产品认证依据产品标准或技术规范和与该产品对应的认证方案；产品检测依据产品标准和方法标准。对于产品认证而言，检测项目、检测方法和判定指标是在符合标准的前提下依据认证方案而定的。也就是说，对于一个建筑产品必须做哪些项目检测是认证机构依据认证方案的要求实施的。而正常委托检测的项目和依据是委托方根据自身的目的和需求向检测机构提出的。因此，产品认证所采信的检测报告项目很可能与正常委托的项目不完全一致。

4. 出具结果不同：产品认证颁发认证证书和认证标志，产品检测出具检测报告。

第三，产品认证并非是质量管理体系认证与产品检测的简单组合，因为证明内容不同。质量管理体系认证没有表明企业生产的产品符合标准的要求。产品检测报告也没有说企业生产的产品和检测样品的关系。只有产品认证是企业生产产品符合标准要求的证明。

第四，不同的认证机构对同一产品的认证可能差别很大。这是我国产品认证的管理制度和市场环境造成的。我国的产品认证就其性质而言，分三类，即强制性产品认证、国家推荐性产品认证和自愿性产品认证。对于强制性产品认证和国家推荐性产品认证，认证方案、检测要求和证书内容以及认证标志是统一规定的，不会产生实质性差别。建筑市场上绝大多数产品是自愿性产品认证，不同认证机构的认证方案肯定是不一样的。不同的认证方案导致了检测项目的多寡、评判指标的高低、抽样或送样的要求、工厂检查的重点和监督的频次等不同。这意味着不同认证机构对同一产品做认证依据各自的认证方案时，出具的结果所包含的信息量很可能大相径庭。

第五，任何一项工作，都会有一个工作结果，无论这个结果是有形的还是无形的。对于产品认证而言，认证证书就是整个认证服务过程的结果。认证证书能够反映很多信息，列举一些主要内容。

1. 申请方的信息，包括申请方的名称和地址。顾名思义，申请方是认证申请的委托方。申请方通常与制造商是一致的，偶尔会是代理商或跨国集团海外的独资或合资的公司。当申请方与制造商不一致时，申请方应有制造商的委托认证的授权。

2. 制造商的信息，包括制造商的名称和地址。制造商是拥有产品知识产权和商标权的公司。

3. 生产厂的信息，包括生产厂的名称和地址。生产厂是产品的实际生产工厂。生产厂可以是制造商或其全资或控股子公司，生产厂也可以是独立法人的生产企业，申请认证的产品仅是其代加工生产的一部分业务工作。对于生产厂信息的披露，我国与欧洲差异很大。根据我国认证证书管理办法的规定，必须在证书上披露生产厂的名称和地址。在欧洲，生产厂可以用生产厂代码表示，而不体现具体生产厂的名称和地址信息。但认证机构知道生产厂的代码指代的具体工厂的信息。

4. 产品信息，包括产品商标、产品型号和产品性能。我国目前的产品认证证书，产品商标和产品型号信息比较齐全。但是对产品性能如何表述，不同认证机构出具证书的差异性很大，现在证书常见的表达形式是产品性能符合标准和认证实施规则要求。这种表达方式的问题在于，一方面大部分认证机构自愿性产品认证实施规则没有公开，另一方面社会潜在采信方不知认证实施规则为何物，因此，认证实施规则对于产品性能的要求对于社会需求方而言是不透明的。此外，符合标准要求的表述也存在信息表达不清的问题。因为，在住建领域标准有产品标准、试验方法标准、设计标准、施工标准、验收标准、技术规程等。对于产品标准，有型式检验项目，即全部检验项目和合格判定指标。对于其他标准和规程可能没有这个条文表述。因此，认证证书上的符合标准要求对于产品性能的表述存在信息量不足的问题。完整的认证证书的信息应使得任何拿到证书的人都清楚知道该产品的哪些具体性能符合哪些标准的要求，性能指标达到多少。

5. 认证评价信息，即认证模式。认证模式是指认证机构对该类产品出具认证证书所采用的评价方法。比如，证书上注明认证模式为"产品型式检验（抽样）+初始工厂检查+获证后监督"，这说明认证机构对该类产品认证进行了工厂检查，进行了抽样，抽的样品依据产品标准进行了型式检验，获证后认证机构对工厂进行了监督。

第二章　产品认证发展

6. 认证技术依据，包括标准或技术规范和认证实施规则。目前对于住建领域所用的建筑产品存在多标准的现象，就是说，对于一个产品可能有多本标准可以采用。至于采用哪本技术标准，取决于认证实施规则的规定。比如，以建筑隔墙用轻质条板为例，目前有3本标准可以适用，分别为《建筑隔墙用保温条板》GB/T 23450—2009、《建筑用轻质隔墙条板》GB/T 23451—2009、《建筑隔墙用轻质条板》JG/T 169—2005。选用上述哪个标准作为认证标准，取决于认证机构自行制定的认证实施规则。在我国的强制性产品认证和国推产品认证中，因为规则是统一制定，所以有助于执行机构依据相同的技术标准实施认证评价工作。

7. 认证有效期和发证机构。认证有效期说明了认证的时效性。目前自愿性产品认证的证书有效期从3年到5年不等，由认证机构自行决定。通过国家认监委的官方网站可以查询到发证机构的资质，有助于社会对发证机构的合规性进行监督。

8. 认可信息。产品认证机构根据国家认监委的授权领域，可以作为认证的标准很多，但并非都在认可范围内。认监委的授权属于行政审批，未经批准私自从事认证或者超出批准范围从事认证活动是违法行为。认可是认证机构的自愿性行为。为了更好地理解认监委的批准书和认可委认可证书的区别，打个比方，认监委的批准书相当于营业执照，认可委的认可证书相当于"能力值得信任的荣誉牌"。对于产品认证领域，认可本应当有两大主要作用，一是增强社会对获得认可的认证机构的信心，二是利用认可实现国际互认。但目前在建设工程领域，这两方面的作用都有待进一步推广宣传，加强认知，推动国际互认工作，从而有助于推动中国标准国际化，助力"一带一路"建设。

第六，不少建设单位的采购部门和主管部门，不了解甚至不知道产品认证为何物，认为只有强制性产品认证才有意义，其实不然。首先，强制性产品认证是国家主管部门发布的，但这个目录是在变更的，目前建筑产品在强制性产品目录中的比例是很小的，未列入强制性产品认证目录的建筑产品并不意味着不重要。其次，企业自愿出资接受第三方认证机构的自愿性产品认证评审，这本身就表明了企业追求诚信，向社会提高自身产品透明度的愿望，也表明了企业对于自己产品质量的信心。再次，强制性产品认证是市场准入门槛，只要从事这个产品生产的企业，都必须通过产品认证，强制性产品认证对于该类产品生产企业没有差异化。

第七，产品认证的起源和发展证明了社会对于产品认证的需求。产品认证的作用是进行专业化的评价，从而向社会发布信息和质量符合公众要求的产品，以便于利益相关方从中进行选择。产品质量是生产出来的，不是认证出来的。这个表述本身没有错误，但以这个表述推出产品认证没有用处的结论就错了。打个比方，健康不是体检保证的，但体检对健康状态进行了有效的监测，有助于健康的改善。

推行产品认证的目的，首先是通过对符合认证标准的产品颁发认证证书和标志，便于利益相关方识别，提高产品性能的透明度，有利于市场诚信体系建设。其次是市场经济发展情况下，国家为了贯彻标准、提高质量、保证安全、保护消费者利益，推行产品认证是政府职能的重要体现之一。再次，产品认证是国际贸易惯例，是进入国际市场的重要条件，推动产品认证和国际互认工作有助于促进国际交流和我国对外贸易的发展。

第三节　我国认证的发展

我国产品认证的发展与我国工业化发展、经济发展，尤其是对外贸易的发展紧密相关。自20世纪70年代始，随着我国改革开放工作的逐步深化，社会主义市场经济建设的发展及加入世界贸易组织，外部和内部环境要求我们去了解产品认证制度。1978年9月，我国以中国标准化协会名义参加国际标准化组织（ISO），成为其正式成员，开始引入认证的概念。经国家有关部门多次组织人员到国外开展认证工作较早的国家和国际标准化组织（ISO）考察和交流，先后翻译出版了一批文件资料，开展了有关认证知识的宣传推广工作，逐步提高了国内政府管理部门和相关人员的认识，基本上统一了在我国开展认证工作的认识。

1981年4月，经国务院标准化行政主管部门批准，我国成立了第一个产品认证机构——中国电子元器件认证委员会。按照国际电子委员会电子元器件质量评定体系的章程和程序规则，组建了认证机构，制定了有关文件，启动了产品认证工作，标志着我国的产品认证试点工作正式开始。

为全面适应我国市场经济发展和加入WTO的工作准备的需要，2001年3月30日国务院进行了机构调整，将原国家质量技术监督局和原国家出入境检验检疫局合并组建国家质量监督检验检疫总局（简称质检总局）、国家认证认可监督管理委员会（简称认监委）和国家标准管理委员会（简称国标委），探索将我国产品的安全认证、强制性监督管理制度与进口产品的安全质量许可制度统一为一个制度，做到"统一标准、技术法规和合格评定程序，统一目录，统一标志，统一收费标准"的强制性产品认证制度。

2001年12月11日，我国正式成为WTO的成员国。同月，国家质检总局发布了《强制性产品认证管理规定》，以强制性产品认证制度替代原来的进口商品安全质量许可制度和电工产品安全认证，强制性产品认证制度在我国正式建立。当时，强制性产品认证管理和实施的主要文件有《强制性产品认证管理规定》、《强制性产品认证标志管理办法》、《第一批实施强制性产品认证的产品目录》、《强制性产品认证实施规则》、《强制性产品认证收费规定》和《强制性产品认证制度实施安排的有关规定》等。强制性产品认证遵循国际认证通行准则，认证制度的建立和运作、认证/检测/检查机构的运作和认证实施规则程序皆遵循ISO/IEC有关国际指南和标准。新的强制性产品认证标志名称为"中国强制认证"（China Compulsory Certification，简称"CCC"标志）。第一批实施强制性产品认证目录涉及19类132种产品，强制性产品认证制度自2002年5月1日起实施，经过1年的过渡期开始强制执行，自2003年5月1日起，所有《目录》内产品须获得强制性产品认证证书，并施加强制性产品认证标志，方可出厂、

第二章　产品认证发展

进口或销售。《目录》中与建筑领域有关的产品为建筑用安全玻璃、瓷质砖、溶剂型木器涂料以及混凝土防冻剂。2014年国家认监委发布第45号《关于发布强制性产品认证目录描述与界定的公告》，共涉及20大类158种产品，其中，与建筑有关产品保持第一批发布的4种产品，未进行调整。

我国在推行强制性产品认证的同时，自愿性认证也在逐步发展。自愿性认证类别很多，目前有质量管理体系认证（QMS）、环境管理体系认证（EMS）、职业健康安全管理体系认证（OHSMS）、危害分析与关键点控制管理体系认证、食品安全管理体系认证、测量管理体系认证、信息安全管理体系认证、软件过程及能力成熟度评估、良好农业规范认证、有机产品认证、食品质量认证、绿色市场认证、无公害农产品认证、绿色食品认证、饲料产品认证、乳制品GMP认证、乳制品HACCP认证，以及其他自愿性产品认证即国家认监委授权认证机构在所有领域的非强制性产品认证，其中包括国推自愿性认证等。国推自愿性认证是指由国家认监委颁布相应的统一认证制度，经批准并具有资质的认证机构按照"统一的认证标准、实施规则和认证程序"开展实施的认证项目。国推认证不是强制性认证，也就是说，在国推认证目录中的产品，不做认证依然可以依法生产和销售，但国推认证具有很强的政府引导性，因此，在政府采购目录以及政府投资项目中，会首先选用通过国推认证的产品。

对于我国认证总体发展情况，根据主管部门的通报，自2009年3月到2014年3月，我国从事认证机构的数量、相应领域颁发的有效证书数量统计情况见表2-2。

不同认证领域的证书统计　　　　表2-2

项目 时间	QMS（张）	EMS（张）	OHSMS（张）	其他产品认证（张）	GB/T 50430（张）	机构数量（个）
2009.03	231420	41317	20935	56693	—	171
2010.03	277314	51806	26908	82412	—	168
2011.03	267927	62964	33296	103330	517	173
2012.06	294367	76746	42040	134312	11303	174
2013.03	306895	86953	52015	167410	19084	174
2014.03	328542	103230	65673	194342	27927	176

质量管理体系是我国开展最早，普及面最广的管理体系认证，截止到2014年3月，有效证书数量为328542张，从2009年3月至2014年3月，有效证书数量的年平均增长率为7.52%。年增长率呈现震荡下行趋势（图2-3、图2-4）。

环境管理体系认证因为近几年对环境问题越来越重视，从法律和监管要求方面越来越多，不少企业开始申请环境管理体系认证。截至2014年3月有效认证证书数量为103230张，从2009年3月至2014年3月，有效证书数量的年平均增长率为20.16%。年增长率呈现震荡下行趋势（图2-5、图2-6）。

职业健康安全同样是近几年我国在法律法规方面明确加强的内容，无论雇主还是雇员对此也越来越重视，截至2014年3月，有效证书数量为65673张，自2009年3月至2014年3月，有效证书数量的年平均增长率为25.7%。年增长率呈现高位平稳

第三节 我国认证的发展

趋势(图 2-7、图 2-8)。

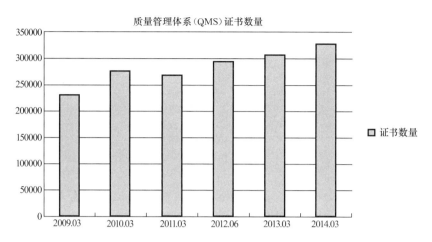

图 2-3 质量管理体系认证证书数量

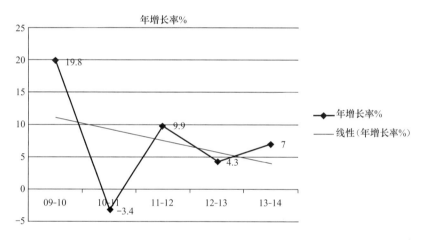

图 2-4 质量管理体系认证证书年增长率

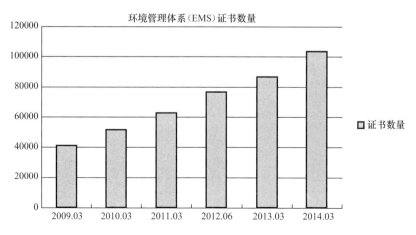

图 2-5 环境管理体系认证证书数量

13

第二章 产品认证发展

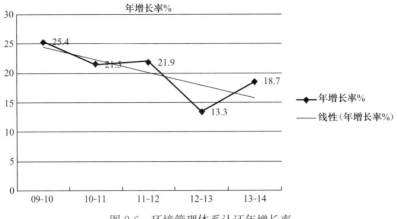

图 2-6 环境管理体系认证年增长率

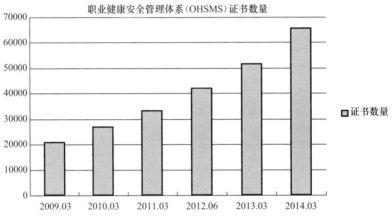

图 2-7 职业健康安全管理体系认证证书数量

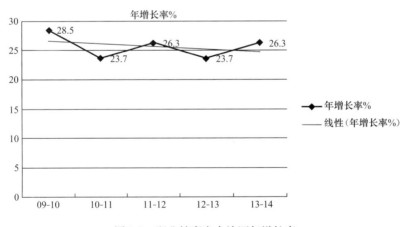

图 2-8 职业健康安全认证年增长率

图 2-9、图 2-10 是我国自愿性产品认证在各个行业所发有效认证证书数量在各年的情况。至 2014 年 3 月有效产品认证证书数量在各个行业的总和为 194342 张。自 2009 年 3 月至 2014 年 3 月，有效证书数量年平均增长率为 28.3%，年增长率呈现下行趋势。

第三节 我国认证的发展

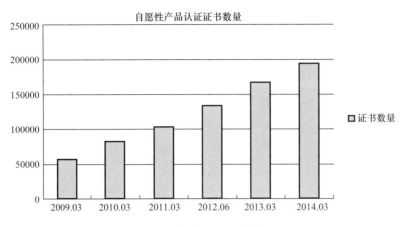

图 2-9 自愿性产品认证证书数量

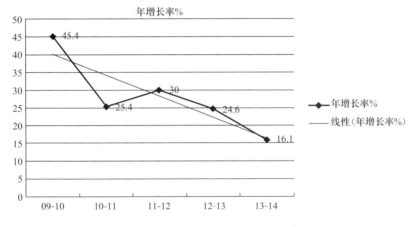

图 2-10 自愿性产品认证年增长率

依据《认证认可检验检测发展"十三五"规划》的指导思想，国家认监委于 2014 年和 2015 年分别针对内资和外资组织申请设立认证机构在行政审批方面进行了重大调整，先后发布了 2014 年第 34 号《国家认监委关于调整在中国（上海）自由贸易试验区内设立外商独资认证机构相关条件的公告》、2014 年第 38 号《关于发布自愿性认证业务分类目录及主要审批条件的公告》和 2015 年第 11 号《国家认监委关于进一步深化认证机构行政审批制度改革有关事项的公告》。这些公告的发布，释放了如下信号：

1. 认证机构资质管理逐步放开；
2. 外商独资认证机构可以进入中国市场；
3. 精简产品认证机构业务领域变更审批程序。

根据最新数据统计情况，仅 2015 年 9 月一个月的时间，新设立认证机构 40 家，截至 2015 年 9 月底认证机构总数已逾 250 家。随着新政策的发布，对认证市场格局必将产生深远的影响，允许民营资本和外商资本进入中国认证市场，强化事中和事后监管力度是"十三五"认证市场的主旋律。

第四节 我国产品认证与政策、法律和标准的关系

随着我国综合国力的增强，国际地位的明显提升，认证作为我国加入国际贸易组织的一项工作，逐步替代了部分原有的行政许可制度，如今已发展成为国家产业规划中的一个重要组成部分。2016年认证认可检验检测已正式作为一个独立的行业依法纳入国家统计局年度行业统计范围。

通过近几届"五年规划"的内容，可以看出国家对认证工作给予了越来越高的重视，发展方向愈来愈明确。在"十五"规划提出"加快电子认证体系和做好加入世界贸易组织的准备和过渡期的各项工作"。在"十一五"规划提出"建立健全电子商务信用和安全认证体系，推行强制性能效标识制度和节能产品认证制度，环境标识和环境认证制度"。在"十二五"规划提出"完善信息安全标准体系和认证认可体系，大力发展认证认可等专业服务，加快低碳技术研发应用，控制工业、建筑等领域温室气体排放，探索建立低碳产品标准、标识和认证制度。推进低碳试点示范。完善能效标识、节能产品认证和节能产品政府强制采购制度"。认证作为高新技术服务业列入了"十二五"大力发展行业。2016年1月，质检总局支树平局长在全国质量监督检验检疫工作会议上用了大量篇幅专门阐述国家质量技术基础（National Quality Infrastructure，以下简称"NQI"）的重要性和紧迫性，科技部也以专栏形式首次将NQI纳入《国家科技创新"十三五"规划》（图2-11）。"国家质量基础（National Quality Infrastructure，简称NQI）"概念是2005年由联合国贸发组织和世界贸易组织首次提出，NQI包括计量、

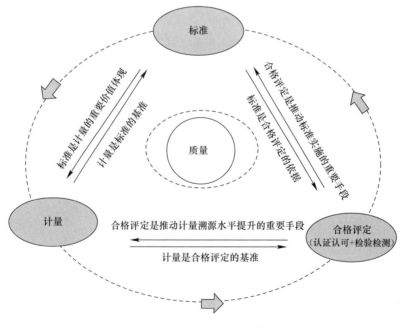

图2-11 国家质量技术基础的概念

第四节 我国产品认证与政策、法律和标准的关系

标准、检测、认证、认可。2006年，联合国工业发展组织（UNIDO）和国际标准化组织（ISO）在总结质量领域100多年实践经验基础上，正式提出计量、标准化、合格评定共同构成国家质量基础，是未来经济可持续发展的三大支柱。

2015年召开的全国认证认可工作会议上，国家认监委提出了六项改革内容，其中包括推进认证市场准入制度改革，适应工商注册"先证后照"向"先照后证"转变，实行机构审批和业务审批"一站式"管理。深化强制性产品认证制度改革，完善产品目录和指定机构动态调整机制，进一步扩大认证检测指定范围，引入企业"自我声明"模式，实行强制性认证证书和标志发放"一站式"服务，提升认证有效性。

为了政策得到有效落实，我国立法机构和行政主管部门制定和修订了一系列与认证有关的法律、行政法规、部门规章、行政规范性文件，见表2-3。

我国与认证有关主要法律法规等文件一览　　　　表2-3

序号	名称	最新发布时间	属性	内容概要
1	标准化法	1988年12月	法律	认证机构授权、产品认证申请
2	产品质量法	2000年7月	法律	认证机构授权、认证机构设立、认证标志使用
3	节约能源法	2007年10月	法律	节能产品认证、政府采购名录采信节能产品认证
4	消防法	2008年10月	法律	强制性消防产品认证
5	标准化法实施条例	1990年4月	行政法规	认证机构的授权和建立
6	认证认可条例	2003年9月	行政法规	规范认证认可活动
7	强制性产品认证管理规定	2001年12月	部门规章	
8	认证证书和认证标志管理办法	2004年6月	部门规章	
9	强制性产品认证机构、检查机构和实验室管理办法	2004年6月	部门规章	
10	认证机构管理办法	2015年8月	部门规章	
11	强制性产品认证标志管理办法	2001	行政规范性文件	
12	强制性产品认证检查员管理办法	2004	行政规范性文件	

认证认可行业标准（以下简称认标）是认证认可技术体系的主要组成部分，是国家标准体系的重要补充和完善。认证的发展与对标准的需求密切相关，标准缺失和不适用的现状影响认证工作的开展。为解决认证认可工作需要，在国家标准委的支持下，认证认可行业标准（标准编号RB）于2012年上半年设立。

认标设立以前，认证活动使用的标准以国家标准和行业标准为主。这些标准并非为认证活动专门定制，很多认证活动由于没有标准或标准不适用导致无法正常开展。虽然国家认监委设立了认证技术规范来解决标准缺失或不适用的问题，但认证技术规范由于知识产权等原因具有很强的排他性，不利于认证行业的发展。认标的制定与发布使认证领域多了一个可选择的标准依据，并且专门为认证活动服务，针对性强。认

第二章 产品认证发展

标使众多规范认证活动开展的技术要求能够标准化，进而指导规范认证活动的开展。

作为国家标准化体系的重要组成部分，认标是认证领域的具体补充和细化，认标与其他行业标准的区别在于，根据 ISO/IEC 17000 附录中 A2.3 规定：应用已有标准或要求时应规定在适当情况下对通用要求进行细化。A2.4 要求在程序、采用方法、人员场所能力等方面可细化规定。以上要求描述了 RB 标准的主要特点。

产品认证在我国的发展要以政策为规划和指导，以法律法规为依据和规定，以标准规范为依托和支持。

第五节 我国绿色建筑产品市场现状

我国在过去 30 余年国民经济持续保持了高速的增长。与此同时，我们也认识到节约资源、能源、保护环境在国家可持续发展战略中体现出来的重要性和必要性。鉴于此，国家政策大力倡导节能、环保、低碳、绿色型产品。那么这些概念指代的是什么产品呢？下面是近些年有关主管部委出台的文件中的表述。

国家发展改革委员会和国家质检总局于 2004 年 8 月 13 日颁布了《能源效率标识管理办法》，能源效率标识是表示用能产品能源效率等级等性能指标的一种信息标识，属于产品符合性标志的范畴。国家对节能潜力大、使用面广的用能产品实行统一的能源效率标识制度。

国务院办公厅于 2007 年 7 月 30 日依据《国务院关于加强节能工作的决定》（国发〔2006〕28 号）和《国务院关于印发节能减排综合性工作方案的通知》（国发〔2007〕15 号），在《国务院办公厅关于建立政府强制采购节能产品制度的通知》中提出节能产品为节能效果显著、性能比较成熟的产品。

环保总局于 2001 年 12 月 23 日颁布了《环境保护产品认定管理办法》。该办法提出，环境保护产品是指用于防治环境污染、改善生态环境、保护自然资源的设备、环境监测专用仪器和相关的药剂、材料。

环境保护部于 2008 年 9 月 27 日颁布了《中国环境标志使用管理办法》。该《办法》提出，在生产、使用及处置等过程中采取一定措施消除污染或减少污染，达到中国环境标志产品技术要求，并通过中国环境标志认证的产品，其生产企业可以向认证机构申请使用中国环境标志。

国家发展改革委员会和国家认监委于 2013 年 2 月 18 日颁布了《低碳产品认证管理暂行办法》。该《办法》提出，低碳产品是指与同类产品或者相同功能的产品相比，碳排放量值符合相关低碳产品评价标准或者技术规范要求的产品。

国家质检总局和国家发展改革委员会于 2015 年颁布了《节能低碳产品认证管理办法》，该《办法》自同年 11 月 1 日起施行。在该《办法》中提出，节能低碳产品认证包括节能产品认证和低碳产品认证。节能产品认证是指由认证机构证明用能产品在能源利用效率方面符合相应国家标准、行业标准或者认证技术规范要求的合格评定活动；

第五节　我国绿色建筑产品市场现状

低碳产品认证是指由认证机构证明产品碳排放量值或者温室气体排放量符合相应低碳产品评价标准或者技术规范要求的合格评定活动。

住房城乡建设部、工业和信息化部于2014年5月21日颁布了《绿色建材评价标识管理办法》，在该《办法》中提出，绿色建材是指在全生命周期内可减少对天然资源消耗和减轻对生态环境影响，具有"节能、减排、安全、便利和可循环"特征的建材产品。

因为上述文件分别发自不同主管部委，所涉及的目录内容会有重叠产品。因此，在建筑产品市场不仅出现了节能、环保、低碳、绿色等概念上的多样化，在市场供需层面也出现了多元化的现象。所谓多元化是指从产品供方提供的证明形式有认证和标识。比如中国质量认证中心出具的太阳能集热器节能认证，新华节水认证中心出具的管材管件的节水认证以及中环联合认证中心出具的中国环境标识，以上三个例子属于产品认证范畴；住建部授权试验室出具的门窗能效标识和国家发展改革委员会会同质检总局以及国家认监委开展的能源效率标识都是主要基于产品检测报告开展的标识评价活动。除上述列举的政府行为以外，市场上还有其他的商业化行为，如商标、推荐书等。

当这种概念多样化和形式多元化同时作用于建筑产品市场的时候，对制造方和采信方都产生了一定程度的信息大爆炸，由于对政府行为和市场行为出现信息甄别困难，对于制造商而言，面对概念多样化和形式多元化，在选择上通常有两种方案，一种是盲从，一种是观望。选择盲从或观望的原则是制造方的利益最大化。当企业发现产品认证是项目准入条件或有鼓励政策时，虽然企业可能不了解产品认证，但也会积极去做，这就是盲从行为。反之，当企业知道产品认证只是自愿行为，做了认证不一定有立竿见影的经济效益时，出于市场认知局限的考虑，企业会选择持续观望态度，直至趋势明朗为止（表2-4）。

企业行为与采信性质矩阵　　　　　表2-4

采信性质 企业行为	准入条件或鼓励政策	自愿行为
盲从	行动	不可能
观望	不可能	行动

对于采信方而言，由于概念多样化、形式多元化，采取的策略是信人不如信己，纷纷建立自己的绿色建材库或绿色建筑产品的采购平台。进入自建数据库或采购平台的方式还是以企业自己供方评价方式为主。之所以产品认证未能在其中发挥传递信任、提升透明度的作用，主要还是源于工程采购方的相关负责人员和采购规则制定的决定人员对产品认证不了解。从另一方面也说明，我国目前产品认证的概念还有待进一步普及推广，而且，不同政府主管部委推出的评价标识和认证政策的查询和咨询方式有待进一步完善。当工程建设方了解产品认证并清楚认证机构专业能力的时候，虽然是自愿性产品认证，更会采信认证结果。反之，不了解产品认证自然很难采信认证结果。但对于强制性产品认证或国推产品认证是例外的，即使采购方不了解，因法律法规规

定或政策导向，也会采信认证结果。但对于自愿性产品认证，由于众多市场因素的影响，即使采购方了解产品认证，也会出现不采信情况（表2-5）。

认证了解度对采信行为的影响　　　　表2-5

采信行为	对认证的了解	了解	不了解
采信		采信认证	适用强制性或国推
不采信		发生在自愿性产品认证	不采信认证

形式多元化的出现说明绿色建筑产品识别评价工作对于市场而言既有需求，又有要求。需求来源于两方面，一是对建筑产品专业能力鉴别的需求，二是政府职能转变的需求。要求来自两方面，一是解决信息不对称的风险问题，二是解决委托代理的风险问题。可以说，解决了市场中信息不对称问题，就有助于满足建筑产品专业能力鉴别的需求；解决了委托代理问题，就有助于满足政府职能转变的需求。

什么是信息不对称问题？具体而言，在绿色建筑产品市场中，卖方（或是生产厂）比买方（招标方和终端消费者）更了解该建筑产品的质量、性能和成本。在交易中，卖方可以通过向买方传递促使其相信的信息而在市场中获益。第三方专业产品认证机构的介入，可以使产品质量信息透明化，从而解决信息不对称问题。

什么是委托代理问题？委托代理理论是制度经济学契约理论的主要组成部分，所指的委托代理关系是一个或多个行为主体根据一种明示或隐含的契约，指定、雇佣另一些行为主体为其服务，同时授予后者一定的决策权利，并根据后者提供的服务数量和质量对其支付相应的报酬。政府委托有资质的机构从事绿色建筑产品的认证，是从社会采购服务的一种具体的形式。在委托代理的关系当中，由于委托人与代理人的效用函数不一样，委托人，比如政府，追求的是履行职能、信息公开、风险可控，而代理人追求的是利益最大化，这必然导致两者之间的利益不一致，甚至由于代理人的非道德行为很可能最终损害委托人的利益，比如在个别案例中，社会上出现的"二政府"和"红顶子中介"问题。绿色建筑产品认证的委托代理的需求是两方面原因产生的，其一是专业化分工要求，其二是市场经济中对政府职能的要求，即"简政放权"。当政府在履行其职能，将绿色建筑产品作为准公共产品处理时，政府如何处理与潜在的有资质的认证机构在授权范围内的委托工作是政策落实成败的重点工作之一，委托代理理论在应用中的核心问题是风险控制。

之所以将建筑产品视为准公共产品，是因为绝大部分建筑产品与日用消费品相比，具有如下特点：

1. 在建造阶段，建筑产品的选择绝大部分情况不是最终用户决定的，而是开发方、设计方和施工方。

2. 最终用户虽然更关心建筑物工程质量、性能和价格，但工程质量、性能和造价与建筑产品的选用密切相关。

3. 对于开发方、设计方和施工方而言，对于建筑产品质量在招采标阶段也很难达到专业化识别。

第五节 我国绿色建筑产品市场现状

上述特点决定了社会个体无法影响建筑产品的质量，而建筑产品的质量关乎着社会成员的生活和健康，乃至社会稳定。鉴于此，为保证工程质量在招采阶段采信有资质的专业机构出具的建筑产品认证结果，不失为一种有效的方式。但是，由于缺乏强制性的规定，招采阶段重价不重质，兼之存在的部分不规范操作，给工程质量隐患埋下了伏笔。

市场对于绿色建筑产品认证的需求和要求恰恰反映了市场经济的部分特征：

1. 绿色建筑产品市场竞争应当是有效的。有效性表现在两个方面，其一是避免地方保护的负面影响，其二是严格制止欺诈、造假等不正当竞争行为。

2. 绿色建筑产品的发展需要"良政治理"，即规范政府职能。市场经济的良性发展离不开政府的作用，政府的作用是在法律约束下促进市场经济发挥应有的资源配置等问题，而不是干预正常的市场行为。因建筑产品与日用消费品相比具有上述特点，所以，建筑产品因按照其涉及安全、健康、节能、环保等内容的实际情况，纳入准公共产品概念。中共中央和国务院 2015 年 9 月印发了《生态文明体制改革总体方案》，《方案》中明确要求建立统一的绿色产品体系，已经表明了这个概念。

3. 绿色建筑产品的发展需要良好的社会信用体系支撑。市场经济由于互联网和大数据的发展，已经从传统的现货交易形式发展为信用交易形式。因此，建立社会化专业化的信用信息服务体系尤为重要。有资质的认证机构在其专业授权范围内可以作为政府的征信机构，解决绿色建筑产品信息不对称问题，促进政务公开，推动社会的和谐发展。

为了更好地解决市场中出现的问题，推动绿色概念的发展和落实，中共中央和国务院于 2015 年 9 月印发了《生态文明体制改革总体方案》，在该《方案》中要求建立统一的绿色产品体系，将目前分头设立的环保、节能、节水、循环、低碳、再生、有机等产品统一整合为绿色产品，建立统一的绿色产品标准、认证、标识等体系。《方案》还要求完善对绿色产品研发生产、运输配送、购买使用的财税金融支持和政府采购等政策。

第三章 我国建筑产品管理

——不积小流，无以成江海

本章主要阐述了如下内容：
- 国家政策：介绍了中共中央和国务院从质量、生态文明建设和国际化等不同方面对建筑性能、产品制造、认证服务提出的要求。
- 产业政策：介绍了发改委、环保部、交通运输部（含原铁道部）、水利部、工信部和住建部开展的节能、环保、低碳、绿色等认证和评价工作。

第一节 国家政策

近几年，中共中央和国务院陆续出台了一系列的政策，从质量、生态文明建设和国际化等不同方面对建筑性能、产品制造、认证服务提出了明确的要求，这些要求也涉及建筑产品领域。

《国务院质量发展纲要（2011—2020年）》提出，建筑工程的耐久性、安全性要普遍增强，节能效率和工业化建造比重要不断提高，尤其是住宅质量性能要明显提高。参照国际通行规则，建立健全法律规范、行政监管、认可约束、行业自律、社会监督相结合的认证认可管理模式，提高强制性产品认证的有效性，推动自愿性产品认证健康有序发展，并稳步推进国际互认。

国务院2015年4月25日在《关于加快推进生态文明建设的意见》中提出，推进节能减排，鼓励新能源利用，建筑工业化；发展循环经济，推进建筑垃圾资源化利用。

国务院在印发《中国制造2025》的通知中提出"质量为先，绿色发展"的基本方针。质量为先包括法规标准体系，质量监管体系，诚信经营市场环境。绿色发展包括循环经济，提高资源回收利用率，构建绿色制造体系。通知中还提出"市场主导，政府引导"的基本原则，明确了建设包括质量认证在内的促进制造业协同创新的公共服务平台。

国务院在《关于推进国际产能和装备制造合作的指导意见》（国发〔2015〕30号）中提出，结合当地市场需求，开展建材行业优势产能国际合作。根据国内产业结构调整的需要，发挥国内行业骨干企业、工程建设企业的作用，在有市场需求、生产能力不足的发展中国家，以投资方式为主，结合设计、工程建设、设备供应等多种方式，建设水泥、平板玻璃、建筑卫生陶瓷、新型建材、新型房屋等生产线，提高所在国工业生产能力，增加当地市场供应。加大中国标准国际化推广力度，推动相关产品认证认可结果互认和采信。

为了落实《生态文明体制改革总体方案》的具体工作，国务院办公厅于2016年11月印发《关于建立统一的绿色产品标准、认证、标识体系的意见》（以下简称《意见》），就贯彻落实《生态文明体制改革总体方案》提出的"建立统一的绿色产品体系"作出部署。《意见》指出，要以供给侧结构性改革为战略基点，坚持统筹兼顾、市场导向、继承创新、共建共享、开放合作的基本原则，充分发挥标准与认证的战略性、基础性、引领性作用，创新生态文明体制机制，增加绿色产品有效供给，引导绿色生产和绿色消费，全面提升绿色发展质量和效益，增强社会公众的获得感。到2020年，初步建立系统科学、开放融合、指标先进、权威统一的绿色产品标准、认证与标识体系，实现一类产品、一个标准、一个清单、一次认证、一个标识的体系整合目标。

《意见》明确了7个方面重点任务。一是统一绿色产品内涵和评价方法，基于全生命周期理念，科学确定绿色产品评价关键阶段、关键指标，建立相应评价方法与指标体系。二是构建统一的绿色产品标准、认证与标识体系，发挥行业主管部门职能作用，建立符合中国国情的绿色产品标准、认证、标识体系。三是实施统一的绿色产品评价标准清单和认证目录，依据标准清单中的标准实施绿色产品认证，避免重复评价。四是创新绿色产品评价标准供给机制，优先选取与消费者吃、穿、住、用、行密切相关的产品，研究制定绿色产品评价标准。五是健全绿色产品认证有效性评估与监督机制，推进绿色产品信用体系建设，运用大数据技术完善绿色产品监管方式，建立指标量化评估机制，公开接受市场检验和社会监督。六是加强技术机构能力和信息平台建设，培育一批绿色产品专业服务机构，建立统一的绿色产品信息平台。七是推动国际合作和互认，积极应对国外绿色壁垒。

《意见》提出了4项保障措施。一是加强部门联动配合，建立绿色产品标准、认证与标识部际协调机制，统筹协调相关政策措施。二是健全绿色产品体系配套政策，加强重要标准研制，建立标准推广和认证采信机制，推行绿色产品领跑者计划和政府绿色采购制度。三是营造绿色产品发展环境，降低制度性交易成本，各有关部门、地方各级政府应结合实际促进绿色产品标准实施、认证结果使用与效果评价，推动绿色产品发展。四是加强绿色产品宣传推广，传播绿色发展理念，引导绿色生活方式。

综上所述，《意见》提出了面对市场上概念多样化，形式多元化的现状的具体解决方法，依据中共中央和国务院规划要求，充分在市场经济中发挥政府职能，全面促进诚信体系建设，对准公共产品通过政府征信的方式，有序推动绿色建筑产品的健康发展，推动生态文明建设。

由上述文件可以看出，我国非常重视建筑领域在生态文明建设中的地位和作用，也同样非常重视产品认证作为质量控制和公共服务以及在国际贸易中所应发挥的作用。对于市场经济中政府职能和信息不对称等问题，文件中也明确提出了"市场主导，政府引导"的原则和促进公共服务平台建设的要求。

第二节 产业政策

建筑产品范围很广，包括但不局限于主体结构产品（如钢筋、混凝土、预制梁、预制剪力墙、预制楼板等）、围护结构产品（如建筑用标准窗、预制墙板、砌块、外保温体系、防水材料等）、装饰装修产品（涂料、装饰板、壁纸等）、暖通空调设备（如热泵机组、太阳能热系统、采暖散热设备、热回收装置、新风系统等）、管线系统产品（如管材管件、阀门等），截至目前，相关部委在落实国务院政策时，都是依据各自行业的特点，推动有关工作。

1. 国家发展改革委员会

国家发展改革委员会在节能和低碳产品领域联合质检总局建立了有关标识和认证的管理办法，并将相关目录中产品纳入政府采购清单。具体实施方案如下：

1）能源效率标识管理办法

国家发展改革委员会和质检总局于2004年联合发布17号令《能源效率标识管理办法》。该办法于2016年2月进行了修订。在该办法中明确了如下内容：

（1）能源效率标识概念：表示用能产品能源效率等级等性能指标的一种信息标识，属于产品符合性标志的范畴。

（2）管理方式：国家建立统一产品目录、统一能效标准、统一实施规则、统一样式；采用备案制度，提交检测报告。

（3）监管模式：地方节能管理部门和地方两局负责监督。

（4）授权实施机构：中国标准化研究院。

（5）收费情况：免费。

（6）配套政策：强制性实施。

2）中国节能产品认证管理办法

原国家经贸委和原国家质量技术监督局在1999年2月颁布了《中国节能产品认证管理办法》。在该办法中提出如下内容：

（1）节能产品概念：是指符合与该种产品有关的质量、安全等方面的标准要求，在社会使用中与同类产品或完成相同功能的产品相比，它的效率或能耗指标相当于国际先进水平或达到接近国际水平的国内先进水平。

（2）评价方式：相关部委制定节能产品目录；采用认证制度。

（3）监管模式：原国家经贸委和原国家技术监督局管理。

（4）授权实施机构：中国节能产品认证中心。

（5）收费情况：收费。

（6）配套政策：自愿性。但为了促进节能产品认证的采信，财政部、国家发展改革委发布了《节能产品政府采购实施意见》（财库［2004］185号）。

3）节能低碳产品认证管理办法

2013年，国家发展改革委员会和国家认监委联合印发《低碳产品认证管理暂行办法》的通知（发改气候［2013］279号），启动低碳产品认证工作。2015年11月质检总局和国家发展改革委员会联合发布了《节能低碳产品认证管理办法》，原《低碳产品认证管理暂行办法》废止。在《节能低碳产品认证管理办法》中，提出如下内容：

（1）概念：节能低碳产品认证，包括节能产品认证和低碳产品认证。节能产品认证是指由认证机构证明用能产品在能源利用效率方面符合相应国家标准、行业标准或者认证技术规范要求的合格评定活动；低碳产品认证是指由认证机构证明产品温室气体排放量符合相应低碳产品评价标准或者技术规范要求的合格评定活动。

（2）评价方式：国家发展改革委、国家质检总局和国家认监委会同国务院有关部门建立节能低碳产品认证部际协调工作机制，共同确定产品认证目录、认证依据、认证结果采信等有关事项。节能、低碳产品认证目录由国家发展改革委、国家质检总局和国家认监委联合发布；采用认证制度。

（3）监管模式：地方两局负责监督。

（4）授权实施机构：节能低碳产品认证机构名录及相关信息经节能低碳产品认证部际协调工作机制研究后，由国家认监委公布。

（5）收费情况：收费。

（6）配套政策：国家发展改革委、国家质检总局、国家认监委以及国务院有关部门，依据《中华人民共和国节约能源法》以及国家相关产业政策规定，在工业、建筑、交通运输、公共机构等领域，推动相关机构开展节能低碳产品认证等服务活动，并采信认证结果。国家发展改革委、国务院其他有关部门以及地方政府主管部门依据相关产业政策，推动节能低碳产品认证活动，鼓励使用获得节能低碳认证的产品。

2. 环境保护部（原国家环保总局）

环境保护部在切实落实国家有关环保政策要求，通过建立中国环境标志和采用产品认证管理模式，积极推动环保工作的开展和环保产品的应用。2003年12月，国家环保总局《关于中国环境标志产品认证工作有关事项的公告》环发［2003］196号；2008年9月，环境保护部颁发"关于发布《中国环境标志使用管理办法》的公告"，进一步明确了环保产品认证工作开展和管理的要求。

1）目的：倡导可持续生产和消费，促进环境友好型社会建设。

2）评价方式：中国环境标志是由环境保护部确认、发布，并经国家工商行政管理总局商标局备案的证明性标识。中国环境标志所有权归环境保护部。环境保护部制定产品目录并负责动态调整。采信产品认证。

3）授权机构：中环联合（北京）认证中心有限公司。

4）收费要求：收费。

5）配套政策：为实现示范效应，环保部与财政部联合落实环保产品的政府采购清单并根据需要更新目录。2015年7月，财政部、环境保护部再次发布《关于调整公布第十六期环境标志产品政府采购清单的通知》（财库［2015］142号）。

第三章 我国建筑产品管理

3. 交通运输部（含原铁道部）

交通部和原铁道部在系统内业务上垂直管理比较密切。在中交和中铁系统建设工程中非常重视产品认证对于工程质量所发挥的重要作用。

交通部先后发布《交通部关于推进交通产品认证工作的意见》（交体法发【2007】738号）和《交通运输行业重点监督管理产品目录（2013年版）》的通知（厅科技字【2013】258号）推动行业内对产品认证的采信工作。目前，交通产品认证机构有2家，分别为中交（北京）产品认证中心和中国船级社质量认证公司。

原铁道部于2003年发布了《铁路产品认证管理办法》（铁科技[2003]104号）。该《管理办法》于2012年修订，由原铁道部和认监委联合发布《铁路产品认证管理办法》（铁科技[2012]95号）。

1）铁路产品范围界定：直接关系铁路运输安全的铁路专用产品。

2）管理主体和内容：国家认证认可监督管理委员会（以下简称国家认监委）负责铁路产品认证工作的监督管理和综合协调工作。铁道部负责铁路产品认证采信工作和认证产品在铁路使用领域的监督管理工作。

3）认证性质：国家对铁路产品认证采取强制性产品认证与自愿性产品认证相结合的方式。实行强制性产品认证管理的，依照国家有关强制性产品认证法律法规的规定执行。实行自愿性产品认证管理的，依照本办法的规定具体实施。

4）采信要求和目录：实行自愿性产品认证管理的铁路产品认证采信目录（以下简称采信目录），由铁道部制定、调整并公布。纳入强制性产品认证管理和列入采信目录的铁路产品，依法取得认证后，方可在铁路领域使用。

5）认证机构：从事铁路产品认证的认证机构（以下简称认证机构）应当依法设立，符合《中华人民共和国认证认可条例》规定的基本条件，具备从事铁路产品认证活动的相关技术能力要求，并符合产品认证机构通用要求的规定。从事强制性产品认证的，还应当经国家认监委指定。从事列入采信目录内产品认证的，还应当经铁道部确认。

6）评价方式：产品认证。

7）授权机构：中铁检验认证中心（原中铁铁路产品认证中心）。

8）收费要求：收费。

9）监管方式：铁道部和认监委联合专项检查。

4. 水利部

水利部依据国务院节约水资源的政策要求，在系统内通过业务垂直管理特点，通过推动节水产品认证重点落实节水工作的开展。2007年水利部办公厅发布《关于加强农业节水灌溉和农村供水产品认证工作的通知》（办农水[2007]144号），各省水利厅转发该通知并在转发中要求辖区内工程采信产品认证结果。2012年水利部、质检总局、全国节水办联合发布《关于加强节水产品质量提升与推广普及工作的指导意见》（水资源〔2012〕407号），该《指导意见》中要求"深入推进节水产品认证工作，建立企业信用档案和产品质量信用信息平台"。

1）管理方式：产品认证。

2) 授权认证机构：北京新华节水产品认证有限公司。

3) 收费要求：收费。

5. 工业和信息化部

工业和信息化部根据所管业务领域范畴，结合认证认可检验检测行业的发展，目前主要开展了电子和通信产品领域的产品认证工作。工业和信息化部通过和国家认监委合作，将与安全、健康、节能、环保有关的产品纳入国家强制性产品认证和国推自愿性产品认证目录。2011年两部委联合发布第18号公告《关于发布国家统一推行的电子信息产品污染控制自愿性认证目录（第一批）和限用物质应用例外要求的公告》。

1) 管理主体：工业和信息化部、国家认监委。

2) 评价方式：主管部委联合颁布产品认证目录；采用产品认证制度。

3) 授权实施机构：泰尔认证中心、赛西认证中心。

4) 收费要求：收费。

6. 住房城乡建设部

住建系统为属地化管理模式，一直以来，侧重点是建筑物在结构安全、节能和绿色方面的性能。在建筑产品领域，市场管理存在多元化的模式，具体说就是生产许可证制度、强制性产品认证制度、地方建筑产品备案制度、产品进场后复检制度与自愿性产品认证制度交错并行。

生产许可证制度主要涉及产品61种，其中仅人造板、建筑用钢筋（冷轧带肋钢筋、钢筋混凝土用热轧带肋钢筋）、预应力混凝土用钢材、建筑钢管脚手架扣件、建筑卷扬机、水泥、建筑防水卷材7种与建筑相关的产品。

强制性产品认证制度仅包括了建筑用安全玻璃（钢化玻璃、夹层玻璃、钢化玻璃）、混凝土防冻剂、瓷质砖、溶剂型木器涂料4类与建筑相关的产品。

上述两种制度涉及管理的产品目录和授权机构由国家质检总局和国家认监委管理。

地方备案制度由各地根据实际情况自行制定，目前主要是各地建委或建设厅所属有关部门负责。产品备案制度是各地各级行业主管部门为加强建筑产品的监督管理所采取的一种管理手段。但是近年来产品备案制度在实施中也暴露出一些值得注意的问题，主要表现为：一些地方在产品备案过程中重复检测、重复发证、重复收费，在实际建筑产品质量管理中没有起到有效监管的作用。

入场复检多数由各地建工领域监督总站负责安排实施。随着我国建筑业的不断发展，现已基本形成了一套比较完善的工程质量管理制度，并制定了大量的技术标准和规范。对建设工程中直接涉及人民生命财产安全、人身健康、环境保护和其他公众利益等方面的要求，以工程建设标准强制性条文的形式加以控制。除建设工程强制性条文外，对影响安全、质量和环境等的建设工程产品质量的控制主要采用进场后的复检制度。但是，由于复检需要一定的检验周期，往往检出后不合格的产品可能已经用于工程，给工程造成很大隐患，并且还需要进一步的实体检测和鉴定，问题严重时，导致工程的拆除，造成人力、物力和财力的浪费。

通过对21个省、自治区、直辖市的建设工程监管机构、检测机构、科研设计院所和建筑

第三章 我国建筑产品管理

产品的生产企业进行建筑产品质量调研,结果显示,目前对于建筑产品承担管理的主要部门依次有各省市工程质量监督站、技术监督局、省建委/建设厅的相关管理部门等(图3-1)。

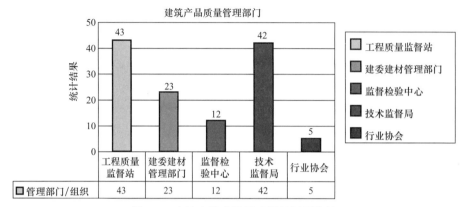

图3-1 建筑产品质量管理部门调研统计

其中,建筑产品的主要监督管理环节为入场环节,由工程质量监督站承担相关的工作。生产和流通环节对于建筑产品质量并未建立有效的控制机制(图3-2)。

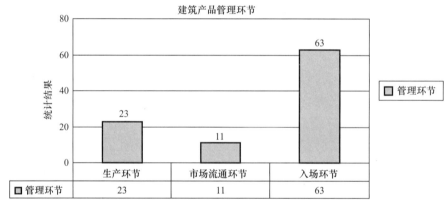

图3-2 建筑产品管理环节调研统计

对于建筑产品的管理主要采用方式依次为入场复检、施工监督和属地备案管理。其中,入场复检是应用最广泛的监管方式(图3-3)。

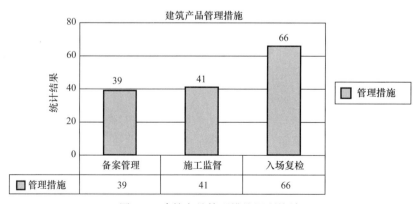

图3-3 建筑产品管理措施调研统计

第二节 产业政策

在备案管理中所查阅的主要资料依次为检测报告、企业资质、质量体系认证证书等文件，采信产品认证的还很少（图3-4）。

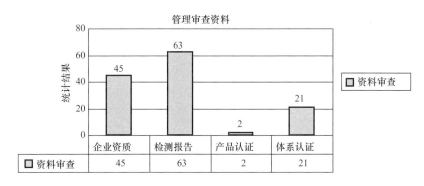

图3-4 资料审查调研统计

通过调研统计，目前影响工程质量的主要因素依次为建筑产品质量问题、施工问题（图3-5）。

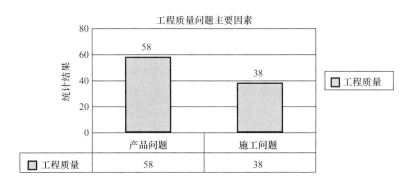

图3-5 工程质量问题主要因素调研统计

通过上述统计，可以得出如下结论：
1）建筑产品质量对于工程质量至关重要；
2）绝大部分建筑产品目前采用入场复检，部分地区辅以属地备案管理；
3）备案管理主要涉及企业三证，部分地区要求提供检测报告；
4）入场复检目前是建筑产品质量管理的主要手段。

问题在于备案管理和入场复检的管理模式不足以从源头控制工程质量。原因如下：

第一，备案管理需要提交的企业三证，是任何合法企业都具有的材料，这些材料与产品质量无关。备案所用的检测报告使用的样品与企业正常生产的产品之间是否一致，缺失验证控制环节。

第二，入场复检和备案时需要提交的检测报告，由于受送样或抽样代表性的影响，从技术角度，存在着送检或抽检样品可能无法客观反映产品批次的质量问题的情况；从采信角度，检测报告仅对来样负责，不对产品质量负责。

其结果是送样或抽样的检测报告不足以客观反映产品的真实质量，导致产品质量对于供需双方的信息不对称，价格成为唯一决定因素。由于供方对于利润空间的追求，

第三章 我国建筑产品管理

降低成本成为必然之路,从而导致产品质量的下降,"劣币逐良币"现象发生,出现市场失灵。

因此,如何构建住建领域质量监管和采信体系,合理运用检测检验和认证制度,是需要探索和解决的问题。

住房城乡建设部于 2010 年发布《关于进一步加强建筑门窗节能性能标识工作的通知》(建科〔2010〕93 号),开启了对建筑产品采用能效标识的管理模式。2013 年由国务院办公厅转发了住房和城乡建设部与发展改革委员会联合发布的《绿色建筑行动方案》,其中提出大力发展绿色建材,并与有关部委研究建立绿色建材认证制度。2014 年 5 月住房和城乡建设部、工业和信息化部联合颁布了《绿色建材评价标识管理办法》。对于绿色建材评价标识,信息如下:

1)管理主体:住房和城乡建设部、工业和信息化部。
2)评价方式:两部委制定产品目录;采用现场评审方式。
3)授权实施机构:由两部委负责三星级评价机构的授权、两部委省级主管部门负责辖区内一、二星级评价机构的授权。
4)收费要求:收费。

7. 小结

相关部委截至 2016 年 12 月底运作模式汇总如表 3-1。

有关部委对产品认证和标识管理一览表 表 3-1

序号	部委	制度名称	运行模式	制度性质	采信方式	是否联合
1	发改委	能效标识、节能低碳认证	非认证、产品认证	强制、国推	政府采购	认监委、财政部
2	环保部	环境标志	产品认证	自愿	政府采购	财政部
3	交通部	认证	产品认证	自愿	系统内招采	否
4	原铁道部	认证	产品认证	自愿/强制	系统内招采/强制	认监委
5	水利部	认证	产品认证	自愿	系统内招采	否
6	工信部	认证	产品认证	国推/强制	招采/强制	认监委
7	住建部	门窗标识、绿色建材评价标识	非认证	自愿	—	否

综上所述,各有关部委在落实中共中央和国务院有关方针政策中,结合了系统自身或垂直管理或分级管理的特点,以及国家财政性资金在计划规模内固定资产投资项目情况,履行政府职能,逐步尝试采购社会服务,保证工程质量,推动生态文明建设,但是构建适宜我国的建筑产品认证制度并与国际工程市场接轨还有待进一步探索和完善。

第四章 国外绿色建筑产品

——博观而约取，厚积而薄发

本章主要阐述了如下内容：
- 选取有代表性国家介绍绿色建筑产品的发展情况

国外代表性国家在绿色建筑和产品领域政策标准概况　　　　　表 4-1

序号	国家	战略、方针、计划	建筑领域政策、标准	建筑产品相关性
1	德国	可持续发展战略	《节能建筑要求》《可持续建筑指南》《节约能源法》《节能法规》	《建筑产品法规》：强制性产品认证
2	英国	可持续发展政策	《建筑法案》《家庭节能法案》《气候变化法案》《环境保护法案》《能源法》	《建设工程的可持续性．环境无害产品声明．产品分类和建筑产品的核心规则》：强制性和自愿性产品认证
3	美国	气候行动计划	"建筑会更好"计划、《能源政策法案》《能源独立和安全法案》《高性能可持续建筑指南》	《可持续建筑技术手册》、GreenFormat、BEES 决策支持系统
4	新加坡	可持续发展战略	《建筑控制法》《建筑控制条例》第五部分"环境可持续性"	自愿性绿色建筑产品认证与绿色建筑评价结合

- 碳足迹：主要介绍了产品碳足迹评价的标准，并分步骤介绍了如何开展碳足迹评价工作。着重阐述了产品类别规则（PCR）在全生命期评价中的作用。
- 环境产品声明：重点介绍了EPD的概念和评价依据，以PVC-U窗和卫生陶瓷为例详细介绍了EPD评价过程。

第一节 德国

1. 政策

1992年，在巴西里约热内卢举办了联合国环境和发展大会，提出可持续发展理念。德国政府据此制定了德国中期和长期发展战略，该可持续发展战略指导方针可总结为"世代公平、生活质量、社会团结、国际责任"，战略主要从三个领域进行落实，分别为经济能力、环境保护和社会责任。德国每四年发布一次可持续战略的进度情况政府报告。德国联邦政府依据可持续发展国家战略出台了"能源气候综合方案"，其中规定与1990年相比，2020年温室气体排放降低40%，在建筑领域，贡献率达到总量的三分之一。

第四章 国外绿色建筑产品

德国是欧洲国家中节能减排法律框架非常完善的国家之一。2004 年德国政府出台了《国家可持续发展战略报告》，其中专门制定了燃料战略，即替代燃料和创新驱动方式。此外，德国政府于 1972 年制定了《废弃物处理法》，该文件于 1986 年进行修改，修改后定名为《废弃物限制及废弃物处理法》。在主要领域开展了一系列实践后，1996 年德国政府颁布了《循环经济与废弃物管理法》。根据德国政府报告，电能消耗的 1/3 都使用在工业和商业部门的用电设备上，因此，2002 年出台了《节约能源法案》，把减少化石能源和废弃物处理提高到发展循环经济的思想高度，并建立了系统配套的法律体系。德国政府于 2007 年 8 月通过了新的保护环境和节能计划，该计划提出了 30 项具体措施，要求德国到 2020 年 CO_2 排放量比 1990 年的水平减少 40%，其中最重要的方法是减少发电，增加建筑的节能和可再生能源的比例。德国非常注重可再生能源的利用，其中《可再生能源法案》的实施是促进可再生能源使用的主要动力。

德国对建筑行业的节能非常重视，不断通过立法提高标准。德国在新建建筑方面制定了严格的使用能源最高限，对保温材料和保温层厚度也制订了严格的标准。在建筑领域为落实具体指标和目标，德国住建系统业务主管部门先后发布了《节能建筑要求》和《可持续建筑指南》等指导性文件，并依据《节约能源法》制定了《节能法规》推动建筑节能工作的开展。《节能法规》2002 年 2 月 1 日生效，同时《保温法规》和《采暖设备法规》废止。《节能法规》自 2002 年首次发布之后，多次修订，最新版是《节能法规》2014 版。《节能法规》修订前控制建筑围护结构的最低保温隔热指标，修订后改为控制建筑的实际能耗。《节能法规》不仅是德国联邦政府能源和气候保护政策的核心组成部分，更是德国建筑经济发展和技术创新的原动力。《节能法规》的颁布和不断修订，一方面逐步落实了德国政府在欧盟对于节能减排的承诺，另一方面促进了被动房和零能耗建筑的推出，带动了建筑围护结构产品和可再生能源设备等创新技术的综合应用。由此可见，对建筑性能的要求会直接影响建筑产品性能的要求。

2. 技术法规

由于德国是欧盟成员国，在执行本国的法律和技术法规时，同时要满足欧盟的要求，在建筑产品领域，影响最直接的是 1998 年德国依据欧盟指令（Construction Product Direction，简称 CPD 指令）制定的《建筑产品条例》。2011 年 4 月 4 日，欧盟发布了《建筑产品法规》，2013 年 7 月 1 日，《建筑产品法规》在欧盟成员国全面实施，至此，1998 年颁布的《建筑产品条例》同时废止。《建筑产品法规》与《建筑产品条例》相比变化如表 4-2。

建筑产品法规调整前后比对　　　　　　　　　　表 4-2

序号	《建筑产品法规》（实施）	《建筑产品条例》（废止）
0	表述：基本要求	表述：必要要求
1	强度、稳定性（未变化）	
2	防火（未变化）	
3	增加：全生命期中温室气体排放、水体污染等	卫生、健康、环保
4	增加：无障碍	使用安全

第一节 德国

续表

序号	《建筑产品法规》（实施）	《建筑产品条例》（废止）
0	表述：基本要求	表述：必要要求
5	噪声控制（未变化）	
6	节能和保温（未变化）	
7	增加：自然资源的可持续使用	无

对于新增加的第 7 条内容"自然资源的可持续使用"在实际应用中转化为如下具体的内容：

1) 建筑拆除后，建筑产品的可回收利用；
2) 建筑产品的耐久性；
3) 建筑产品的环保性能；
4) 从全生命期进行评价。

作为欧盟成员国，德国依法转化为德国《建筑产品法》和《建筑法规》。《建筑产品法》和《建筑法规》中都明确规定了如下内容：建筑产品一致性证明方法，一致性认证，检测、检查和认证机构的要求。技术细节的要求在标准规范中体现。

《建筑产品法规》是强制性执行，所涉及的建筑产品依据其在安全、健康、节能和环保等方面的重要性差异，分别执行产品认证或自我宣称方式。同样由于产品重要性不同，采用不同的产品认证制度。

3. 标准体系和认证体系

作为法律法规的技术支撑，德国的标准规范体系主要由两部分组成，即欧盟协调标准（hEN）和欧洲技术许可指南（ETAG）。2015 年欧洲技术许可指南（ETAG）已开始逐步过渡为欧洲评估文件（EAD）。

欧盟协调标准（hEN），由欧盟标准化委员会（CEN）负责归口管理。如果产品在欧盟协调标准范围内，且属于欧盟强制性产品认证（CE）范围，必须执行该标准。欧盟协调标准的制定既要满足欧盟相关法律法规的基本要求，又要考虑到各国应用和转化的要求。

欧洲技术许可指南（ETAG）和欧洲评估文件（EAD），由欧洲技术评估组织（EOTA）负责归口管理。符合指南要求后由授权机构出具欧洲技术评估（ETA）报告，生产企业通过技术评估后，获得 CE 认证标志，用于证明建筑产品性能符合工程使用的需要（图 4-1）。

由于建筑产品的多样性，既有成熟标准产品，也有大量非标产品，还有针对工程项目的专有产品。对于成熟标准产品，欧盟协调标准已经覆盖，对于非标产品，欧洲技术许可组织（EOTA）编制了欧洲技术许可指南（ETAG）和欧洲评估文件（EAD），如混凝土用锚栓、外墙外保温体系、预制楼梯、结构胶等。对于特定工程的专有产品，则依据《欧盟建筑产品条例》9.2 款进行单独评价。

在 CE 认证中，对于不同产品的风险和特征进行了分类，纳入不同的体系等级。对于不同的体系等级，采取的评价模式不同（表 4-3）。

第四章 国外绿色建筑产品

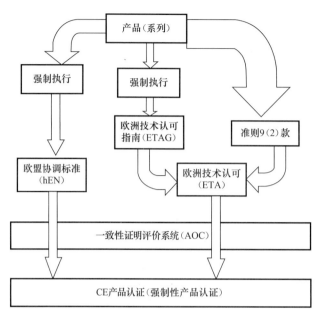

图 4-1 欧盟 CE 产品认证框架

欧盟 CE 产品认证一致性证明体系（Attestation of Conformity，简称 AOC） 表 4-3

体系等级	授权机构的任务	生产商的任务
体系 1（1+）	初次型式检验 见证测试 针对生产控制的初次工厂检查 对于生产控制的持续监督、评价和批准	生产控制 抽样测试
体系 2	针对生产控制的初次工厂检查 对于生产控制的持续监督、评价和批准	初次型式试验 生产控制 抽样测试（执行规定的测试方法）
体系 3	初次型式试验	生产控制 抽样测试（执行规定的测试方法）
体系 4	—	初次型式试验 生产控制 抽样测试（执行规定的测试方法）

对于不同的体系等级，从社会征采第三方机构的服务也不相同（表 4-4）。

欧盟 CE 产品认证不同体系等级与第三方机构的关系 表 4-4

体系	涉及的第三方机构
体系 1（1+）	认证机构、工厂检查机构、检测机构
体系 2	认证机构、工厂检查机构
体系 3	检测机构
体系 4	/

因为建筑产品的多样性和欧盟区内各国环境、地理以及气候性的差异，德国一方面需要遵守欧盟的法规条例，另一方面需要确保本国的工程质量。因此，德国依据

第一节 德国

《建筑产品法》和《建筑法规》制定了建筑产品目录。在欧盟 CE 强制性产品认证范围内的建筑产品，执行 CE 认证。在 CE 认证范围之外，但在德国建筑产品目录内的产品，必须执行德国 Ü 产品认证（图 4-2、图 4-3）。

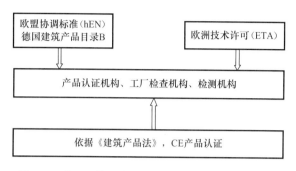

图 4-2 欧盟及德国境内 CE 产品认证的依据和机构

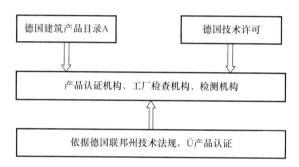

图 4-3 德国 Ü 产品认证的依据和机构

德国在建筑产品领域实施的 Ü 产品认证分为 ÜHP 和 ÜZ 两种模式，这两种模式所对应的一致性证明体系和涉及的第三方机构见表 4-5、表 4-6。

德国 Ü 产品认证一致性证明体系 表 4-5

体系名称	授权机构的任务	生产商的任务
ÜHP	初次型式检验（工厂抽样）	生产控制 抽样测试
ÜZ	初次型式检验 见证测试 针对生产控制的初次工厂检查 对于生产控制的持续监督、评价和批准	生产控制 抽样测试

德国 Ü 产品认证不同体系与第三方机构的关系 表 4-6

体系	涉及的第三方机构
ÜHP	检测机构
ÜZ	认证机构、工厂检查机构

由此可以看出德国对于建筑产品的管理有如下特点：
1) 依据欧盟法规条例，结合本国实际情况，制定产品管理目录；
2) 根据产品安全等级等特性，明确产品所属的管理等级；

3) 明确授权机构和授权范围；

4) 采信产品认证的结果。

对于欧洲技术评估（ETA），成员国都仅有1个机构作为授权机构。德国的授权机构为德国建筑技术研究院。对于欧洲协调标准（hEN），涉及认证机构、工厂检查机构和检测机构。德国境内这些机构都是由德国建筑技术研究院依法指定，并通知欧盟管理机构备案并公示。目前，根据市场需求和机构自身技术能力，德国境内授权依据欧洲协调标准（hEN）认证的机构有19家，涉及不同的产品标准，基本没有重叠业务领域，且认证机构几乎同时都是检测机构和检查机构，检查机构19家，涉及不同的检查领域，个别检查机构同时也是检测机构，仅从事检测的授权检测机构8家，所有机构都通过了德国认可委（DAAK）的认可。

4. 监管体系

欧盟作为统一市场，遵循4个原则，即自由贸易、自由旅行、自由提供服务、自由金融流通。为了确保上述原则得以落实并有效运行，2008年7月9日欧洲议会制定了条例（EC）No.765/2008，规定了市场监管的要求。

市场监管的范围涉及市场上自由流通的CE产品认证。监管内容主要包括：

1) 建筑产品是否符合CPD指令要求；

2) CE标识的使用是否符合条例（EC）No.765/2008的规定；

3) 核查建筑产品是否存在严重的风险。

德国各联邦州设有市场监察机构，负责各自属地内涉及建筑产品的质量监管工作。德国建筑技术研究院负责协调各州监察机构的行动，并在需要时提供技术支持。德国建筑技术研究院作为州市场监察机构的代表出席欧洲有关建筑产品市场监察的会议并参与承担其中的主要工作（图4-4）。

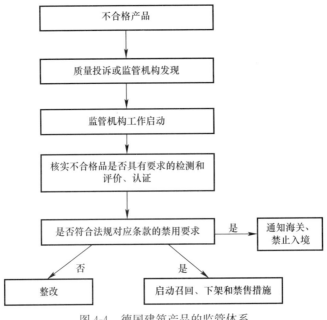

图4-4 德国建筑产品的监管体系

对于市场监管，德国采用主动监管和市场反应相结合的原则。主动监管是监察机构对工程项目实施核查的工作。市场反应是指当出现主动监管外的质量问题时所采取的措施。

第二节 英国

1990年英国政府发布的《环境白皮书》明确把可持续发展列为今后建设必须遵循的国家战略。1994年，英国积极响应里约全球环境首脑会议要求，率先制定可持续发展战略《可持续发展：英国的战略选择》，为英国的节能环保发展建立了一个基础性文件。1999年5月，英国政府公布第二份可持续发展战略，确定了同步发展经济、社会和环境的目标，并引入了量化指标。2005年3月，英国政府出台第三份可持续发展战略，规划了到2020年英国社会和环境的发展方向。2007年7月，英国政府发布新的可持续发展指标统计，详细地介绍了68个领域的发展变化状况。

英国政府以5年规划为基础，建立了国家可持续发展战略。英国的可持续发展政策包括五方面内容：第一，生活在有限资源下，敬畏地球资源，合理利用，留泽后代；第二，确保坚强、健康和公正的社会，满足人民需求，提供人民富裕水平，为所有人提供平等的机会；第三，实现经济可持续发展，构建强大、稳定和可持续发展的经济，为所有人提供机会和前景，同时降低每人承担的社会成本；第四，应用科学的理念，确保政策的发展基于科学的基础并考虑其不确定性，同时尊重公众的态度和价值观；第五，提升良政治理，提升政府管理体系的有效性和参与性，激活公众对此的创造性、活力和多样性。

英国可持续发展战略在国家政治经济生活各个领域都予以落实和体现，是英国政策平衡发展的支柱。英国在落实可持续发展战略过程中有如下特点：

1) 内阁的领导和检察；
2) 政府的表率作用，如政府建立并使用绿色采购平台，从行政角度监督政府办公建筑和其他公共建筑的节能；
3) 在各领域落实可持续发展战略，使绿色概念融入英国家庭的日常生活；
4) 确保透明性和独立性。

英国建立了绿色经济体系，发展绿色交易，制定碳价格，建立绿色发展基金。绿色交易在建筑领域主要体现在与建筑相关法规、建筑控制制度和节能性能认证等方面。为了降低建筑能耗和节约能源，英国政府在强制性的建筑节能法规（Document L: Conservation of fuel and power）里规定了建筑的最低节能要求，法规给出建筑节能的指导意见和参考标准，并且此系列标准每4~5年都会修订一次，每次都会提出更新、更高的标准。英国在节能建筑中采取的技术措施主要有三个方面：一是采用构造措施，提高墙体、屋面及门窗等围护结构的保温性能；二是利用太阳能等可再生能源；三是改进供热系统。

第四章　国外绿色建筑产品

截至 2015 年底，英国还是欧盟成员，从立法层面，英国建筑法规和建筑产品法规遵守欧洲法规规定，如《欧盟建筑能源性能指令》。2016 年英国举行公投脱离欧盟后，英国结合国情出台或修订相关法案，如《建筑法案》《家庭节能法案》《气候变化法案》《环境保护法案》《能源法》等。在操作层面，对于不同建筑分别采用 BREEAM、Eco-Home、Sustainable Home 评价体系。对于建筑产品，有强制性标准的，采用强制性认证，其他采用基于 EN15804《建设工程的可持续性．环境无害产品声明．产品分类和建筑产品的核心规则》的自愿性产品认证。只有通过指定认证的产品才能在上述建筑评价体系中获得加分。

第三节　美国

1. 政策、法规和标准

美国政府在 20 世纪 70 年代末 80 年代初因能源危机问题，开始制定并实施建筑物的能源效率标准，先后出台《能源政策和节能法案》《节能政策法和能源税法》《国家能源管理改进法》。1992 年制定了《国家能源政策法》并于 2003 年进行了修订，将以前的"目标"转换为"要求"，实现了节能标准从规范性要求到强制性要求的转变。1992 年里约热内卢联合国环境和发展大会后，美国政策开始注意到节能对气候和环境的影响，对新能源和可再生能源的激励措施越来越多。1998 年公布了《国家能源综合战略》。2001 年针对新建节能建筑颁布税收激励政策。2005 年颁布的《能源政策法案》是目前美国实施绿色建筑、建筑节能的法律依据之一，对于提高能源利用效率、更有效地节约能源起到了至关重要的作用，标志着美国正式确立了面向 21 世纪的长期能源政策。奥巴马就任美国总统后，启动了"气候行动计划"（Climate action plan），作为"气候行动计划"的重要组成部分，"更好的建筑"（Better building）计划在 2011 年启动。在此计划中，阐述了美国联邦政府和各州政府的政策方针、金融机制、劳动力市场和建筑节能信息化等内容。此外，美国《节能与节能产品法令》要求美国能源部在 IECC（国际节能规范）的住宅节能规定和 ANSI/ASHERA/IESNA90.1 标准修订版出版一年内审定新版本是否提高能效。一旦美国能源部确定修订版提高了能效标准，要求各联邦州在 2 年内评估检查各自住宅和公共建筑节能法规修订的必要性。

在最低能效标准方面，制定了 IECC（国际节能规范）2000 标准和 ASHRAE（美国采暖制冷与空调工程师协会）标准，对低层住宅、商用建筑和高层建筑能源性能（包括围护结构和采暖空调），如最小热阻值和最大传热系数等方面作了强制性要求。

从节能建筑的内容看，美国从两方面提出建筑节能要求，一是建筑物本身的热工性能，即通过提高建筑围护结构的保温隔热性能、门窗的密闭性能和充分利用通风、自然采光等措施来减少采暖和空调能源的能耗；二是提高建筑内部的能耗系统及设备的能源效率，包括采暖、空调、照明、热水器、家用电器及办公设备等。每年由美国能源部主办的建筑节能法规大会是美国全国性的重要活动之一，大会主要议题包括对节能法规的培训、研讨法规和标准的修订方向与办法等。

2. 绿色建筑产品与建筑

在《能源政策法案》、《能源独立和安全法案》、联邦政府关于《高性能可持续建筑备忘录》等法律法规和行政规章的基础上，美国联邦政府发布了《高性能可持续建筑指南》。美国绿色建筑委员会结合能源主管部门和环境保护主管部门的有关文件，发布了《可持续建筑技术手册》。在《手册》中阐述了建筑产品的选取和产品认证要求，从而在绿色建筑设计阶段明确提出建筑性能对于建筑产品的要求，将建筑产品的评价和绿色建筑评价有效结合。

现在设计师和建筑商越来越多地被要求使用"绿色"建筑产品。究竟如何实现对绿色建筑产品的选择？如果产品具有可回收性，那么产品是否更属于环境友好型呢？单一性能评价、多性能评价，以及全生命评价，产品在建筑中发挥的绿色度如何计算？主流产品总是不如市场上被认为是"环保型"的产品优先？做绿色产品肯定更贵吗？

为了解决上述的问题，BEES 在美国由 NIST（国家标准和技术研究所）、美国环境保护署（EPA）、建筑火灾研究实验室在美国环保优先采购（EPP）计划和白宫赞助的住房技术推广伙伴关系（PATH）的支持下开发。在政策执行层面，EPP 计划负责执行联邦政府通过废物预防、回收和采购的绿化行政命令，鼓励机构减少与他们每年购买的 2000 亿美元的产品和服务相关的环境负担，其中包括建筑产品。BEES 正在进一步开发，作为协助联邦采购团体执行 13101 号行政命令的工具。在技术开发层面，美国国家标准与技术研究所（NIST）与美国环境保护署（EPA）国家风险管理研究实验室空气污染防治控制部于 1997 年联合制定规范的研究方法和公开的数据库，用于平衡建筑材料的环境性能和经济性能。EPA 负责开发环境绩效数据数据库，并且在 EPA 支持下，由 NIST 负责将经济性能数据添加到数据库，开发决策支持软件，用以实施设定的 LCA 算法以平衡建筑材料的环境和经济性能，并在建筑设计师和材料制造商中推广实施。组合的软件和数据库产品将被称为 BEES（Building for Environmental and Economic Sustainability，环境和经济可持续发展建筑），该决策支持软件将实现对环境和经济绩效数据库的访问，结合具体的产品进行运算。该决策系统将使用环境和经济性能数据的数据库作为输入。BEES 将通过互联网提供即时访问工具并对传输数据实现即时优化。如可以随着环境评估现状的变化，新建筑材料到达现场的情况，以及建筑材料的成本改变信息，测算随着时间的推移，预期数据的改进情况。BEES 可以适应不同水平用户的专业知识和需求。它包括内置的"默认"数据，以便不熟悉 LCA 的用户可以轻松实现建筑材料选择。但是，BEES 不包括环境和经济绩效的相对重要性的默认值。相反，BEES 将显示决策备选方案提供的环境和经济利益。它将由用户选择最能反映其观点的替代物。经验丰富的用户将能够自定义默认数据。例如，材料制造商可以对其产品输入专有数据。其他数据，如环境影响的相对重要性权重和 LCA

计算的折现率，都是可编辑的。因此，这些用户将能够进行"假设"分析，以检查如何改变数据而影响环境与经济绩效平衡。此外，BEES 将遵循 LCA 方法的数据透明度原则记录每个 LCA 阶段使用的数据和假设条件。

BEES（环境和经济可持续发展建筑）为用户提供了选择具有成本效益的"绿色"建筑产品的技术。该决策支持系统是基于标准而设计，具有实用性、灵活性和透明性。BEES 决策支持系统面向设计师、建设者、产品制造商、科研人员、咨询机构和监管机关等利益相关方。BEES2.0 版包括超过 65 个通用建筑产品的实际环境和经济性能数据。

BEES 使用国际标准化和基于科学的全生命期评估方法测量建筑产品的环境性能，分析产品生命中的所有阶段：原材料获取、制造、运输、安装、使用、回收和废物管理。在整个全生命期阶段测量多达十种环境影响，即全球变暖、酸雨、资源枯竭、室内空气质量、固体废物、富营养化（土壤和水中不必要的矿物营养物的增加）、生态毒性、人类毒性、臭氧消耗和烟雾等。BEES 采用全面的多维统计测算，该方法的亮点在于强调了真正减少总体环境影响必须作出的权衡。

BEES 将全生命期的思维方式应用于测量经济绩效。经济绩效使用 ASTM 标准全生命期成本测量法，涵盖初始投资、更换、运营与操作、维护和修理以及废弃处置的成本。全生命期成本法在固定时间段内将这些成本相加，称为研究期。用于相同功能的替代产品，可以根据其全生命期成本进行比较，以确定哪个产品是在研究期间实现最低成本的手段。为了将环境和经济性能结合到整体性能测量中，BEES 使用 ASTM 标准进行多属性决策分析。BEES 用户可以指定用于组合环境和经济性能指标的相对重要性权重，并且可以测试总分数对不同组合的相对重要性权重的敏感性。

那么，如何使用 BEES 来比较竞争产品的环境和经济性能？让我们通过例子说明。假设我们考虑两种地毯，一种为使用常规胶水安装的宽幅织物尼龙地毯，另一种主流替代品是由 PET（再循环软饮料瓶）制成的宽幅地毯，并使用低 VOC 胶（相对低水平挥发性的有机化合物胶）安装。

第一步是使用图 4-5 所示的 BEES 窗口设置分析参数。

如果我们不希望将环境和经济绩效指标合并，我们可以选择"不加权"选项，这样仍然计算分解的 BEES 结果。否则，我们需要设置重要性权重。在本例中，环境绩效和经济绩效具有同等重要性，因此两者都设置为 50%。接下来，我们需要为包含在 BEES 环境绩效分数中的环境影响类别设置相对重要性权重。我们选择"等权重"，即指定对所有影响同等重要处理。我们的最后一个参数是用于将未来建筑产品成本转换为等价现值的真实贴现率。在这里，采用接受默认率 4.2%，该数值是美国管理和预算办公室对大多数联邦项目的规定。接下来，我们需要为每个地面覆盖替代品设置最后一个参数——从制造设施到将要安装产品的建筑工地的运输距离。此参数允许 BEES 计算使用本地生产产品的重要性的环境绩效值。如图 4-6 所示，我们选择了尼龙地毯替代品的运输距离 805 公里（500 英里）。

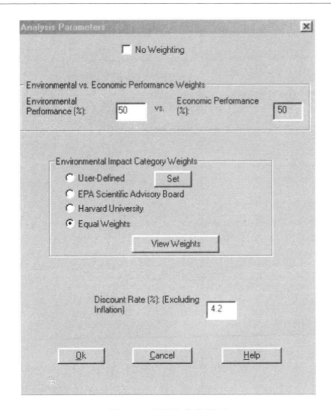

图 4-5　BEES 参数设置

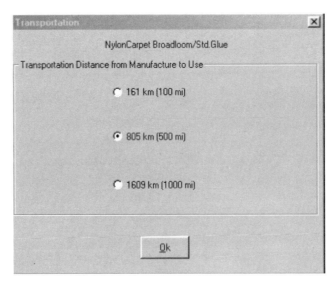

图 4-6　运输里程设置

现在我们准备好计算和查看 BEES 结果。图 4-7 以图形和表格形式显示示例的加权环境绩效分数结果，数值越低越好。如果产品在所有环境影响方面表现较差，则得到最差的 100 分。在本例中，尼龙宽幅地毯的总分为 96 分，PET 宽幅地毯的总分为 49 分。

第四章　国外绿色建筑产品

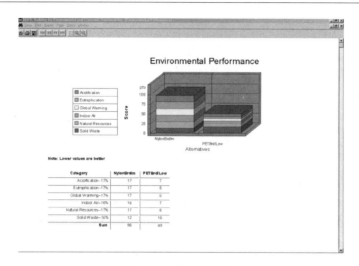

图 4-7　环境绩效分数

该图通过其对酸化、富营养化、全球变暖、室内空气、自然资源枯竭和固体废物六个方面分解加权环境得分。如图所示，PET 地毯在除固体废物之外的所有影响类别上表现更好。根据显示，可以看到每个影响类别旁边是其分配的相对重要性权重。

图 4-8 显示了本例中的经济绩效结果，它给出了第一成本、贴现未来成本，以及它们的总和，即全生命期成本。该图显示 PET 宽幅地毯具有较高的全生命期成本（每 0.09 平方米安装地毯的现值美元为 10.21 美元，或每平方英尺 10.21 美元，而尼龙为 4.57 美元），具有较高的首次成本和较高的未来成本，较高的未来成本是由于更高和更频繁的重置成本造成的。

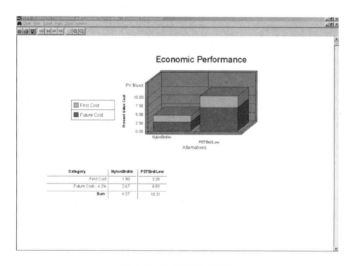

图 4-8　经济绩效结果

因此，根据我们指定的折扣率 4.2%（显示在未来成本类别旁边的表格中），PET 宽幅地毯的分数更好，而尼龙宽幅地毯的经济性更好。

整体性能分数为我们提供了一种结合和平衡环境与经济性能分数的方法。图 4-9

显示了基于我们对环境和经济绩效的相等权重的 BEES 总体绩效结果。它显示每个产品替代方案的总体绩效得分，这是其加权环境和经济绩效得分的总和。表中每个性能类别旁边，是其分配的相对重要性权重。

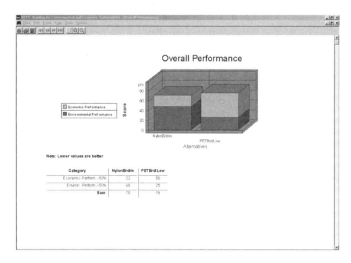

图 4-9　环境绩效和经济绩效总分

从该图可以看出，尼龙宽幅地毯的得分为 70 分，PET 宽幅地毯的得分为 75 分。因此，基于我们的分析参数，安装有常规胶的尼龙宽幅地毯总体上略微优于安装有低 VOC 胶的 PET 宽幅地毯。请注意，除了此处显示的摘要图表之外，BEES 还提供了每个环境影响的详细图表（例如，报告每个产品对全球变暖影响的二氧化碳克数），这有助于确定产品全生命期中"薄弱环节"对环境的影响。

将 BEES 方法应用于 BEES 2.0 决策支持系统中包括的建筑领域产品有框架、外墙和内墙饰面、墙体和屋顶护套、顶棚和墙体保温、屋顶和地板覆盖物、板、地下室墙壁、梁、柱、停车场铺路和车道等。通过测算得出如下结论：

第一，基于单一影响的环境要求，例如单纯的全球变暖，应该持怀疑态度。这些碳排放数据没有考虑到一个影响可能以牺牲其他影响为代价而改进的事实。

第二，评估必须在功能单元的基础上量化，因为它们在 BEES 中，使得被比较的产品是彼此间可以实现真正的替代。以单位质量为基础时，一个屋顶覆盖产品可以在环境上优于另一个，但是如果该产品需要两倍的质量作为另一个产品的替代品，以覆盖 1 平方米的屋顶，则结果可能相反。

第三，产品可以包含显著影响成分，但是如果该成分是相对良性产品的一小部分，则其显著性会明显降低。不可因噎废食。

第四，短期内，低成本产品通常不是成本效益高的替代品。对于耐用的、免维护的产品，较高的第一成本可以接受。

BEES 将在未来几年内在如下方面进行扩大和完善。首先，将更多的产品添加到系统中，以便可以比较整个建筑部件和系统。为此，鼓励建筑产品制造商通过新的 BEES 申请计划提交特定的性能数据。其次，更多的环境影响，如居住地改变等因素将会纳入未来版本的 BEES。最后，引入美国区域特异性等因素，测算会有更大的灵活性。总而言之，BEES 的答案在于权衡，决策支持系统是预测建筑对环境有关的问题与相关对应的成本效益减少的结果。

3. 绿色建筑产品信息标准化

GreenFormat 是由建筑规范研究所（CSI）成员开发的一种信息组织工具，用于传达建筑材料、产品和系统的可持续性特征。CSI 建议行业利益相关方，无论是信息提供者，如建筑产品制造商，还是信息用户，如建筑业主、设计师、总包方或承包商，遵循 GreenFormat 来构建这种信息。CSI 的 GreenFormat 是一种标准化结构格式，通过使用这种标准化的格式，制造商可以准确地识别产品的关键特性，并向设计方、建造方和施工方提供满足可持续要求所需的信息。通过使用 GreenFormat 收集和组织的信息可用于生成产品的可持续发展概况。使用 GreenFormat 识别标准和适用的认证结果为设计方、建造方和施工方提供了一种简便的方法来评估各种制造商的材料和产品的可持续性。

虽然对"绿色"或"可持续性"没有精确的定义，但许多专家的共识是，由于所有人类活动都伴随着对环境的影响，所以可持续性是照顾人类现在的需要，同时不危害环境，避免对后代造成不利影响。它涉及将我们生活的环境、经济和社会等方面与我们居住的地球如何结合的方式。在建筑领域主要涉及集成设计方法，以便将现场、能源、水、材料、室内环境和社会使用建筑物的方式等其他方面要素综合考虑。理论上讲，这些要素必须予以考虑，以确保建筑环境的最佳性能，同时限制其产生的影响。

当前市场，"绿色"和"可持续性"已成为非常混乱的术语，现在有众多利益相关方参与这一主题，有理论家、设计从业人员、认证机构和商业实体，通过产品标识或认证宣称匹配绿色建筑要求。还有不同的评级系统，旨在通过改善能源性能和其他方式来减少建筑物的环境影响，例如绿色建筑倡议中的 LEED® 和 GreenGlobes®，还有其他评级系统包括美国能源部的能源之星计划和 ASHRAE 189。关键问题是，在可持续发展方面有许多指南、评级系统、法规和标准，信息作为一种特殊产品在市场供需上出现混乱。

不同的利益相关方在寻找可持续发展相关信息时，他们的目标可能是不同的，对于同一指标的优先级也不一样。比如，制造商对于"绿色建筑产品"的自我声明适用吗？当行业专家与制造商对于绿色建筑产品的理解不一致时，应该考虑吗？使用单一属性标准，如回收内容，而不是全面的多属性全生命期评估数据如何？如何让需要信息的人，无论他们是设计从业者、建筑业主、材料供应商还是其他建筑行业专业人士，根据他们最重要的需求来决定，这是 GreenFormat 方法设计的初衷。GreenFormat 旨在为与可持续发展相关的各种信息提供类别。提供者获得他们认为重要和相关的信息，并且该信息的用户决定如何使用它并且作出自己的选择。

GreenFormat 最初被用作 CSI 推荐的格式，用于制造商的产品信息报告，于 2006 年首次发布在 USGBC。用户很快发现它作为产品比较工具很有用，这促成了 2008 年发布的 GreenFormat 数据库。虽然该计划取得了一些成功，各种复杂性和保持网站内

第三节 美国

容更新的成本使得 CSI 考虑采取新的方法。因此，在 2012 年 CSI 开始从基于网络的产品选择工具转变为综合产品信息格式化工具。CSI 的目标是支持由行业专业人士使用 GreenFormat 作为组织、查询和使用产品可持续发展相关信息的最佳方式。

重要的是 GreenFormat 不是一个特定的信息产品，而是一种广泛的产品信息工具可以应用的格式。这可能包括但不限于产品搜索和比较工具、网页、数据库、产品数据表、演示工具，以及可能的许多其他工具。它的体例格式用以对应于 CSI 的 SectionFormat 的第二部分和第三部分，组织规范信息的标准。当设计专业人员编写规范时，他们包括许多"显著特点"作为项目要求。这些突出特征中的每一个具有与其相关联的所需值。例如，产品的典型单属性特征可以是"再循环内容"，并且说明符合要求的"值"可以是"50%"。GreenFormat 包含许多适用于产品可持续性方面的突出特性，然后制造商可以提供适用于自己产品的特征值，说明可以将这些特征值合并到他们指定的产品的第二部分描述中。有关产品安装和维护的信息可以包括在第三部分中。GreenFormat 是制造商出版产品信息的有用指南。正如 MasterFormat 为工作结果提供了统一的层次结构和顺序一样，GreenFormat 为产品信息提供了统一的层次结构和顺序。它满足用户对产品数据评估、数据归档和产品比较的需求。

为什么设计专业人员会使用 GreenFormat？该组人员通常由建筑师、工程师和室内设计师组成。有一点需要注意的是，他们通常有很多选择。他们可以选择替代技术或替代系统，如不同材料、其他类型的产品等；换句话说，选用不同的方式来建造事物。由于所有人类活动都会影响环境，设计团队的目标通常是选择按预期执行的产品，以可接受的成本实现总体环境影响最低。GreenFormat 可以帮助实现这一点。这个团体可以节省时间和金钱来获取关于可持续产品和材料的信息。他们可以坚持他们的设计意图，选择更容易实现的产品选型决策，这也同时增加了客户的满意度。GreenFormat 可以帮助项目更快完成，并在设计、规格和施工阶段更好地控制。通过使用 GreenFormat 支持产品审议和选择过程，可能减少风险责任，因此信息管理也会令人头痛。使用 GreenFormat 可以改善和增强行业的沟通。简单来说，GreenFormat 可以帮助减少设计对环境的影响。

那么，建筑产品的供应商如何从使用 GreenFormat 中受益？供应商由制造商、销售代表、经销商、经销商、行业协会和其他与产品和材料相关的实体组成。他们通常寻求推广他们的产品，并影响设计的选择和评价性标准规范的制修订过程。他们的目标是通过支持他们的决策过程，并在市场上有效地竞争，从而游说设计师并影响产品选型。GreenFormat 使设计专业人员更容易找到产品，从他们将考虑的许多替代品中选择产品，并通过指定它来使用产品。GreenFormat 是为了更好的沟通而设计的。所有建筑产品的制造商都喜欢保持他们的客户满意度，并鼓励他们回来再次购买。通过使用 GreenFormat，他们可以发挥作用，阐述他们的产品对建筑环境的影响。现在，制造商可以开始组织所有与产品可持续性相关的信息，并根据 CSI 的 GreenFormat 向消费者提供信息。指定的受众对 CSI 有信心，他们认识到清晰、高效的沟通的重要性，使他们能作出最好的选择。帮助他们更好地完成工作的一个重要方法是使用 Green-

Format作为信息管理工具来指定产品。现在，设计专业人员可以要求供应商根据CSI的GreenFormat结构提供产品信息。制造商对设计师非常感兴趣，他们喜欢知道如何最好地与他们沟通。他们将喜欢指导如何给设计专业人员提供他们需要的最佳方式。

总之，GreenFormat解决了建筑行业众多利益相关者的信息需求，包括设计专业人士（建筑师、工程师、设计师和室内设计师）、供应商（制造商、经销商和分销商）和其他影响者，包括建筑业主、建筑商和监管机构。

4. 产品认证

随着绿色建筑的发展和对绿色产品的需求，美国出现了一些专门从事建筑行业产品认证的实体。有两个很活跃的组织，绿色印章（Green Seal）和科学认证体系（SCS），他们基于自己的指南建立了产品的评价标准。自定的评价标准虽然没有遵循一个完整和开放的共识过程，如美国材料试验协会（ASTM）的做法，这两个实体还是征求了不同程度的公众意见。一旦制定了标准，Green Seal和SCS就会审查制造商的产品信息。如果产品信息及相关规格和性能数据超过标准，产品将获得组织的认证和相关标志。Green Seal是一家致力于环境标准制定、产品认证和公共教育的非营利组织。其目的是在技术和经济可行的范围内减少与产品的制造、使用和处置相关的环境影响。美国保险商实验室公司（UL）是Green Seal的主要检测和检验承包商。绿色认证产品超过50类，并已授予其近240种产品，包括一些建筑行业产品，如紧凑型荧光灯、高效夹具、油漆、窗户、窗膜、热水器、密封剂、胶粘剂和抗腐蚀涂料等。除了识别以环保方式设计和制造的产品外，Green Seal还提供科学分析，帮助消费者作出关于环境影响的有针对性的购买决策，并鼓励制造商开发环保产品。SCS是一家私营科学组织，其使命是促进私营和公共部门实现更环保的可持续发展政策、产品设计、管理和生产。SCS制定了与可持续建筑实践相关的工作，如环境声明认证，这是一种用于验证产品环境声明准确性的系统；生命周期评估主要是协助组织进行全生命期评估，制定改进的战略和记录成效；生态认证是提供产品及其包装的环境概况。

第四节　新加坡

新加坡国土面积很小，自然资源非常匮乏，政府忧患意识很重。因此，新加坡政府的可持续发展战略理念发展相对较早，认识程度较高。2005年新加坡政府对国内化石燃料消耗进行了统计，其中，工业、交通、建筑所占比例分别为31.7%、1.58%、31.9%。建筑领域对于化石燃料的消耗占了将近三分之一的比重。因此，新加坡政府大力推动绿色建筑，以期降低对于化石燃料的消耗。新加坡的绿色建筑倡导在确保建筑环境质量和舒适度都不会降低的情况下实现低能耗和环保。

新加坡政策法规规定公共服务部门在新建建筑和既有建筑改造方面起表率作用，应符合绿色建筑技术要求。国家所属建筑均强制性率先达到绿色建筑评价入门级标准。这方面要求和英国一样。

从技术法规方面，新加坡建设局修订了《建筑控制法》，自 2008 年 4 月 15 日实施新《建筑控制条例》，规定了"可持续标准"的最低限度。新的《建筑控制条例》由五部分组成，其中，第五部分就是《建筑控制（环境可持续性）条例》，要求新建建筑和进行重大改造的既有建筑至少达到规定的最低环境可持续性标准，即达到绿色标志认证合格等级，它表明所有新建建筑及部分既有建筑改造被纳入强制性认证的范畴。

从市场运作方面，新加坡的模式和美国相似，建立了政策奖励机制，包括绿色标志建筑面积（GM-GFA）鼓励计划和既有建筑绿色标识激励计划（GMIS-EB）。奖励方式有两种。其一，对于达到绿色标志白金和黄金＋等级的开发项目，将在总体规划的总容积率控制线之外额外许可其一定的建筑面积（黄金＋项目额外许可 1% 的建筑面积，对于白金项目额外许可高达 2% 的建筑面积）。其二，对有建筑所有权企业建造了高等级绿色标志建筑，在这些企业进行住宅的运营、租赁业务时，政府把税点降低，使他们获得较高的单价运营利润。

为提升建筑业技术方面的能力，政府鼓励企业项目的开发与协作，包括回收和循环使用绿色建筑材料，以及有效节省材料的设计，优化自然资源的利用。新加坡开展了绿色建筑产品认证，如使用具有回收性质的材料，就可以获得绿色标志，从而使绿色建筑产品和绿色建筑有机结合起来。但也可以看出，目前新加坡所开展的绿色建筑产品认证的评价指标侧重于不同产品单一方面（如节能、环保、可循环）的性能要求，不具备全生命周期的特点。

第五节 碳足迹

1. 碳足迹的标准

随着全球气候变暖，温室气体的作用被重视，碳足迹的概念孕育而生。碳足迹源于英语"Carbon Footprint"，这个术语是指具体产品在其全生命期内的各种 GHG 排放，即从原材料获取到生产制造、销售、运营使用和废弃后处理等所有阶段。温室气体（GHG）范围引用 IPCC 2007 中的温室气体范围，包括二氧化碳（CO_2）、甲烷（CH_4）、氮氧化物（N_2O）、氢氟碳化物（HFC）和全氟碳化物（PFC）等温室气体。碳足迹是因消耗能源而产生的二氧化碳排放对环境影响的指标。

在对产品的温室气体评价方面，《PAS 2050—2008 产品与服务生命周期温室气体评估规范》是最早提出评价产品全生命期内温室气体排放的规范，它由碳基金和英国环境、食品和乡村事务部（Defra）联合发起，英国标准协会（BSI）编制。ISO 14067 是在 PAS 2050 的基础上发展而来的。

评价任何一类产品在其全生命期内的 GHG 排放可以包括两种类型，第一种是从商业到消费者的各类商品，这种情况客户是指终端用户，即 B2C 模式；第二种是商业到商业的各类商品，这时客户是另一个商户，而该商户将该产品用作其自身各种活动的输入，即 B2B 模式。对于建筑产品而言，除部分装饰装修材料和家用暖通空调产品外,绝大部分属于B2B模式。

不同企业生产同一类建筑产品,只有产品温室气体排放统计评价采用统一的数据源、边界条件、统计方法和其他假设条件,统计评价的结果才有可比性和意义。

对于具体的产品,确定功能单位是必要的一步。为了进行碳足迹计算,功能单位可以被认为是某一特定产品的一个有意义的数量。根据 PAS 2050 和 ISO 14067 的表述,功能单位是用作基准单位的量化的产品系统性能,一个功能单位反映了产品被最终用户实际消费的方式,如 1 立方混凝土、1 吨 PU 的保温材料等。

ISO 14067 和 PAS 2050 概括强调了应用 ISO 14040 和 ISO 14044 规定的 LCA 方法学进行评价的原则。PAS 2050 提出相关性、完整性、一致性、准确性以及透明度五个原则:

1)相关性是指选择适合于评价所选产品生命周期 GHG 排放的源、数据和方法。

2)完整性是指包括所有对一个产品生命周期排放提供"实质性"贡献的 GHG 排放和存储。

3)一致性是指在 GHG 的相关信息中能够进行有意义的比较。

4)准确性只是尽可能地减少误差和不确定性。

5)透明度是指在通报结果时,披露足够的信息,以允许第三方作决定。

ISO 14067 不仅包含上述五个原则,还对迭代计算方法、科学方法选择顺序等作了补充规定。

2. 评价碳足迹步骤

对产品进行温室气体排放统计评价过程,通常包括如下环节。

步骤 1:过程图绘制

绘制产品全生命期过程图,从原材料获取、生产制造、运营使用到废弃后处理,包括所有的材料流和能量流。这一步骤的目的是确定对所选产品全生命期内有显著贡献的主要材料、活动和过程。对于绝大部分建筑产品,依据 ISO 14040 采用 B2B 模式而言,只包括从原材料通过生产直到产品到达一个新的组织,包括分销和运输到客户所在地。它不包括额外的生产步骤、最终的产品分销、零售、消费者使用以及废弃后处理。

步骤 2:边界确认

系统边界定义了产品碳足迹计算的范围,即依据产品过程图哪些阶段的输入和输出宜纳入评估。系统边界的关键原则是列入对产品全生命期主要有实质性的排放,即选定原材料获取、生产制造、运营使用和废弃处理等过程中直接或间接产生的排放。对于实质性贡献 PAS 2050 和 ISO 14067 有所不同,PAS 2050 对于实质性贡献明确提出超过该产品生命周期预期排放总量实质性贡献是指超过该产品生命周期预期排放总量 1%的任一来源的贡献。但是,非实质性排放源的总的比例不得超过整个产品碳足迹的 5%。

ISO 14067 和 PAS 2050 以及在我国市场上常见的环境标志产品认证的依据是不同的。ISO 14067 依据 ISO 14025,即Ⅲ型;PAS 2050 依据 ISO 14021,即Ⅱ型;我国现行的环境标志产品认证依据 ISO 14024，即Ⅰ型。三种不同的类型有着不同的命名，Ⅰ型叫环境标志，Ⅱ型叫自我环境声明，Ⅲ型叫环境产品声明（即 EPD）（图 4-10）。由于三种环境标志采用的评价方法不同，实施起来有着巨大的区别。Ⅰ型的特点是对

CPC	Name
	Insulation materials
	Flexible sheets for waterproofing - bitumen, plastic or rubber sheets for roof waterproofing
	Stand-alone construction product PCRs, NOT compliant with EN 15804
	Stand-alone construction product PCRs, compliant with EN 15804
	Sanitary ware of iron, steel, copper or aluminium (under development; replacing PCR 2012:10)
	Windows and doors
1533	Asphalt mixtures (under development)
264	Synthetic carpet yarn used for construction purposes (sub-oriented PCR; appendix to PCR 2012:01)
27922	Nonwovens for other purposes than clothing (expired 2016-09-19)
31600	Builders' joinery and carpentry of wood (expired 2014-12-19)
3511	Paints and varnishes and related products
37129	Acoustical systems solutions (sub-oriented PCR; appendix to PCR 2012:01) - previously Acoustic ceilings
3731	Bricks, blocks, tiles, flagstone of clay and siliceous earths (sub-oriented PCR; appendix to PCR 2012:01)
37330	Mortars applied to a surface (sub-oriented PCR; appendix to PCR 2012:01)
3744	Cement
375	Concrete
42911	Sanitary ware of iron, steel, copper or aluminium (being replaced)
4299	Fabricated products made out of metal composite material (MCM)
54	PCR 2012:01 Construction products and construction services (combined PCR & PCR Basic Module)

图 4-10　EPD® 采用的建筑产品类别规则目录

第四章 国外绿色建筑产品

每类产品制定产品环境特性标准，在我国，就是 HJ 标准，如对于涂料，HJ 仅规定有害物质限制的指标。Ⅱ型是企业自我环境声明，这种模式在我国随着诚信体系建设完善会逐步普及。Ⅲ型才是全生命周期评价。

对于产品全生命期的系统边界依据 ISO 14025 宜有对应的产品种类规则（简称 PCR）。产品种类规则（PCR）根据 ISO 14040 标准，对一个或多个产品种类进行Ⅲ型环境声明所必须满足的一套具体的的规则、要求和指南。PCR 提供了一套一致的、国际公认的方法，可用于定义产品全生命期。目前，建筑领域产品的 PCR 数量仍然有限。图是国际组织目前已发布的建筑产品的 PCR 清单。

步骤 3：数据收集

PAS 2050 根据 ISO 14044：2006，4.2.3.4.3 的数据质量要求，将评价指标划分为初级活动水平数据和二次数据。初级数据是指针对具体产品生命周期由内部或者是由供应链中别人所做的直接测量；次级数据是指不针对具体产品的外部测量，但是一种对同类过程或材料的平均或通用测量。PAS 2050 要求初级活动水平数据应用于所有过程和材料，即产生碳足迹的组织所拥有、所经营或所控制的过程和材料。一般情况下，尽可能多地使用初级活动水平数据，因为这类数据可使人更好地了解实际排放情况，并有助于找到提高效率的真正机会。凡无法获得初级活动水平数据或者初级活动水平数据质量有问题（例如没有相应的测量仪表）时，有必要使用直接测量以外其他来源的次级数据。在某些情况下，只要可行，为了确保一致性并具有可比性，次级数据可能更为可取。碳足迹数据质量的评价主要是准确性、重现性及可比性三方面，同时对地理上、距离上和材料上的不同也予以考虑。

ISO 14067 引用 ISO 14044：2006，4.2.3.6.2 的数据质量要求，除以上几方面考虑还要求检验数据的代表性和不确定性，并提出进行碳足迹研究的组织应具有数据管理系统。

步骤 4：碳足迹计算

计算碳足迹需要两类数据：活动水平数据和排放因子。活动水平数据是指产品生命周期中涉及的所有材料和能源（物料输入和输出、能源使用、运输等）。排放因子是一种联系，可将这些数量转换成温室气体排放量，即单位活动水平数据排放的温室气体数量，比如每千克输入量或每千瓦时能源使用量的千克温室气体。活动水平数据和排放因子可来自初级或次级数据。

计算碳足迹以质量守恒为基础，通常质量平衡测算，确保所有输入、输出和废物流均被计入。质量平衡测算的目的是确认所有材料已被全部计入，没有任何物质流的重大遗漏。在 100 年评价期开始时，不能将那些因长期使用（比如外保温材料）或在最终处置阶段随时间推移产生的排放物视为排放物的一次性释放。因此，必须对这些排放量进行计算，以体现评价期内排放物在大气中的加权平均时间。之所以有 100 年的时间概念测算，源于全球增温潜势（GWP）这一术语。GWP 是用来描述一种温室气体单位相对于一个二氧化碳当量单位在 100 年内产生的影响。二氧化碳当量是指衡量所有温室气体的全球增温潜势的一个计量单位。因此，产品碳足迹的公式是整个产品生命周期中所有活动的所有材料、能源和废物乘以其排放因子之和。计算本身只是将相应排放因子与活动水平数据相乘即可。

某一活动的碳足迹＝活动水平数据（质量/体积/千瓦时/千米）×排放因子（每个单位的CO_2当量）。

为确保温室气体数据的有效性、权威性和可比性，排放因子应由国家主管部委在调研统计并论证确认后统一发布。活动水平数据应由根据Ⅲ型还是Ⅱ型的情况，由认证机构或企业进行统计并实施核查。活动水平数据统计和排放因子的采用宜考虑到如下影响因素。

第一，供应链的变更。由于各种原因，生产企业的供应链可能发生变化，这会影响统计数据的准确度。

第二，产品最终的处理形式。计算产品的碳存储需要理解被测算产品在100年内的处理形式。如某些产品可能被焚化，某些产品可能被填埋，某些产品将被再生利用。依据PAS 2050的测算方法，当采用焚化处理时，需要识别甲烷是否被收集并用于发电，若产生有用的能源，排放不包括在该产品的碳足迹中并被分配给所生产的能源部分，反之，则无能源回收，化石碳（并非植物碳）产生的排放包括在该产品的碳足迹中（如同填埋）。当采用填埋处理时，废物中植物碳的CO_2排放不包括在内，即植物碳排放的GWP值被赋予0，如采用植物纤维制成的轻质墙板；非化石碳产生的CO_2排放包括在产品碳足迹中，其GWP值被赋予1。当一种产品可以被再生利用时，该产品的碳存储效益即终止。但是使用再生利用材料的产品可获得碳存储效益。用于计算可回收利用的产品的碳排放量方法取决于原材料本身以及原材料的回收利用系统是否是产品系统本身的一部分。

第三，能源相关排放可能来自于燃料燃烧、发电或供热。

1）对于现场生产并使用的能源，应从初级活动水平数据中计算排放因子；

2）对于场外生产的能源可使用供应商或其他可靠的原始来源提供的排放因子；

3）对于可再生电力可使用可再生电力特定排放因子；

4）对于生物质燃料的CO_2排放，若燃料是从废物中生产获得，相关的排放量即为废物转化为燃料过程中造成的排放；若燃料是从植物物质生产获得，则排放量包括生产和使用该燃料产生的整个生命周期的排放量。

第四，统计数据的分配。对产品的全生命期在运输过程产生的GHG排放量应纳入碳足迹评估。当该产品连同其他产品一并运输时，运输产生的各种排放则根据物理质量或体积进行分配。在生产过程形成共生产品或副产品时，则需要对排放进行分配。对于共生产品或副产品，可以理解为非废物，并且具有经济价值，可以出售。

步骤5：不确定性检查

对产品碳足迹的不确定性分析是一种对精度的衡量，用于确定某个输入或计算的准确度或精确度。通过对不确定性来源的分析，通常可以降低数据的不确定性。如采用质量好的初级活动水平数据替代次级数据，或者采用质量更好的次级数据，改进用于计算碳足迹的模型等，都是行之有效的方法。当然，并非所有的初级数据的不确定性一定低于次级数据,但是对于某个特定过程或排放源,用不确定性评估来判定采用初级数据还是次级数据是一种很好的判定方式。

第六节 环境产品声明 EPD

1. 环境产品声明概念

环境产品声明(Environmental Product Declaration,简称 EPD),这份声明是由独立的第三方机构核实并对声明中指定的产品在全生命期内造成的环境影响的报告。该报告的信息对于相关方是透明的,报告中评估的数据具有可比性。需要说明的是,一款产品出示 EPD 声明报告并不说明该款产品就一定优于其他产品,而是说明该款产品给社会相关方一份公开的、透明的在全生命期内对环境影响的说明。各相关方可以根据声明中数据信息选择适合的产品用于适合的项目。在推动建材产品向绿色建材发展过程中,先后出现了环境标志、碳足迹和 EPD 的概念。这三者是有区别的。

首先,这三个概念的依据不同。环境标志依据的主要标准是 ISO 14024 和方案制定者编制的有关评价标准规范或规则;碳足迹评估依据的主要标准是 PAS 2050 或 ISO 14067 以及产品类别规则(Product Category Rule,简称 PCR);EPD 声明主要依据标准是 ISO 14025、ISO 14067 和 PCR。PCR 是针对不同产品分别给出的全生命期的评价文件,PCR 文件通常包括特定产品全生命期边界划分、各阶段包括评估内容、数据处理中舍去和分配原则等规定。

其次,这三个概念评估产品的内容不同。环境标志根据相关技术要求关注产品环保性能;碳足迹是从全生命期评估产品的温室气体排放;EPD 是从全生命期评估产品的性能、能源和资源的使用,以及由此造成对环境因素的影响,包括全球变暖潜势(Global Warming Potential,简称 GWP)、酸化潜势(Acidification Potential,简称 AP)、富营养化潜势(Eutrophication Potential,简称 EP)、臭氧层损耗潜势(Ozone Depletion Potential,简称 ODP)、光化学臭氧合成潜势(Photochemical Ozone Creation Potential,简称 POCP)等。

因此,EPD 评估对于产品而言是一份全面的评估。

2. EPD 运营体系

欧洲建筑产品环境声明的推出与实施是各方共同努力的结果。这包括标准层面、实施层面和采信层面三大部分。

标准层面主要是由国际标准(ISO)、欧洲标准(EN)和产品种类规则(Product Category Rule,简称 PCR)构成。国际标准主要涉及 ISO 14025 和 ISO 14044,欧洲标准主要是 EN15804,产品种类规则是根据具体产品编制详细的评价内容。

实施层面主要是由检测机构、认证机构和评价机构共同完成。首先,由认证机构根据认证要求制定检测方案,安排检测机构实施产品检测。认证机构实施工厂生产控制(Factory Production Control,简称 FPC)的检查。认证机构负责汇总检测报告和工厂检查报告,颁发认证证书,确保产品性能符合法律和标准规范的要求。在此基础上,评价机构依据上述 ISO 14025、EN 15804 和 PCR 标准体系的要求,进行环境声明的评价,并最终出示 EPD 报告。

采信层面主要是政府的推动工作。大多数建筑产品,比如主体结构预制构件、围护结构预制板等,都涉及人体安全和健康等问题,而大多数建筑产品恰恰不是最终使用方能够决定的。因此,建筑产品具有公共产品的属性。对于公共产品依据市场经济特点,政府就必然行使其"良政治理"的职能。大部分建筑产品在满足强制性产品认证的前提下,政府会依据国家战略发展方针引导公众接受 EPD 概念(图 4-11)。

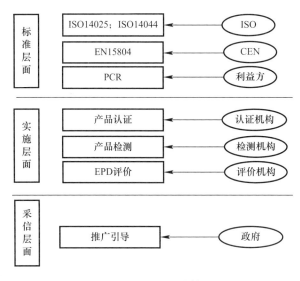

图 4-11　EPD 运营体系

3. EPD 评估示例——PVC-U 窗

下面以 PVC-U 窗为例,说明 EPD 评估的步骤和实现方法。

步骤一:确认产品评估系统边界和内容

为进行 PVC-U 窗的 EPD 评估,须要确认 PCR 文件依据是否存在。这里依据 PCR 2012:01 和基于 ISO 14040 以及 ISO 14044 的全生命期评价原则,确定产品评估边界和评估内容。对于 PVC-U 窗,全生命期评估主要由三个阶段组成,分别为制造阶段、使用阶段与废弃处理阶段。

制造阶段包括原材料的组成分析、生产制造工艺分析统计。使用阶段包括从生产厂运输至施工地点、现场安装、使用年限内零部件更换,以及使用年限内的能源补偿。废弃处理阶段包括 PVC-U 窗和零部件的拆除、回收再利用,以及废弃处理。

步骤二:产品描述

本例为 PVC-U 窗,产品基本信息如表 4-7。

这里使用年限分别考虑了 10 年、30 年以及 50 年三种不同的情形。

步骤三:生产用材料信息

该款 PVC-U 窗的组成为 PVC-U 窗框(包括密封条)、平板玻璃及五金件。因为该规格窗尺寸适中，窗框没有设置钢衬。各部分材料的质量组成信息如表 4-8。

PVC-U 窗的产品信息　　　　　　　　　　表 4-7

序号	项目	PVC-U 窗
1	尺寸	1 米×1.3 米
2	窗框厚度	70 毫米
3	表面积	1.3 平方米
4	使用年限	30 年（10 年，50 年）
5	玻璃	双玻
6	保温性能 K	1.4
7	采光性能 T	0.5

PVC-U 窗材料构成信息　　　　　　　　　　表 4-8

序号	项目	PVC-U 窗（1 米×1.3 米）	
		质量（千克）	比例（%）
1	窗框（PVC-U）	12.7	39
2	密封材料（密封条）	0.5	1
3	五金件（钢材）	2.2	7
4	平板玻璃	17.2	53
5	总数	32.6	100

窗框是由 PVC 粉末和添加剂通过挤出工艺生产完成。其中，PVC 粉末是主要原材料，添加剂包括改性剂、填料、色母和热稳定剂。这些添加剂将保证通过挤出加工工艺生产的型材性能能够满足设计要求。

步骤四：分析原材料对人体健康和环境的影响（表 4-9）

原材料有害性分析　　　　　　　　　　表 4-9

序号	材料	比例（%）	环境影响	人体健康影响
1	PVC	82	/	/
2	稳定剂（CaZn）	3	/	吞咽有毒、对眼部有严重伤害、可能引发皮肤过敏
3	色母（T_iO_2）	3	/	/
4	填料（碳酸钙）	7	/	/
5	改性剂	5	/	/

上述材料影响评价的依据主要是《危险化学品目录（2015 版）》，这版是根据联合国《全球化学品统一分类和标签制度》（Globally Harmonized System of Classification and Labelling of Chemicals，简称 GHS）制定的。

步骤五：使用阶段

使用阶段评估包括 PVC-U 窗运输到工地，安装，更换窗户零部件和考虑能源损失补偿。理论上，PVC-U 窗的使用年限可以为 50 年，但我国实际使用年限通常在 10 年左右。所以，本例中在评估测算时分别采用了 10 年、30 年和 50 年不同使用年限情况。

PVC-U窗在使用期间，建筑物室内和室外因为温差的存在时刻进行着热交换。为了保证建筑内使用者的舒适度，将根据所在地区不同、季节不同，采取供暖或制冷的措施进行补偿。这就是能源补偿产生的原因。

由此产生的能源补偿多少取决于很多因素，包括地区气候条件、建筑物的采光朝向、使用者的生活习惯等。上述因素很难精确评估测算，但恰恰这些因素会导致很大的测算偏倚。因此，在处理使用阶段能源补偿时，通常采取平均值测算，并务必给出评估测算所依据的假设条件，这些假设条件包括但不局限于：

1) 地理位置因素，比如属于我国哪个热工区域，甚至可以进一步细化。
2) 暖通空调系统运行采用平均值，比如包括拟采用的供暖和/或制冷的设备信息。
3) 鉴于目前PVC-U窗气密性能较好，由PVC-U窗密封不严而导致的气流引起的热交换损失忽略不计。
4) 通风而产生的热损失不予考虑。

在上述假设的条件下，使用期间能源补偿需求为使用PVC-U窗造成的热交换损失和冬季太阳能透射获得热能之差。

步骤六：废弃处理阶段

建筑物废弃后，PVC-U窗不同部件将进行回收再利用或废弃处理。建筑垃圾的废弃处理通常是由专业处理公司完成。因为目前国内尚未发布官方的建筑垃圾分类回收率的情况，本例的测算依据欧洲有关协会发布的数据。PVC型材能够实现约为50%的回收率，其中，密封条和型材的涂层视为与型材共同处理。五金件属于钢材，钢材回收率比较高，主要是考虑可以通过电磁吸引分离技术实现回收，世界范围内平均回收率约为80%。平板玻璃回收再利用率依据欧洲发布的数据约为15%。

步骤七：环境影响因素

环境影响因素主要包括五方面：全球变暖潜势（GWP）、酸化潜势（AP）、臭氧耗减潜势（ODP）、光化学臭氧合成潜势（POCP）、富营养化潜势（EP）。评估时将把全生命期各阶段的测算数据汇总，依据数据取舍和分配原则核算到产品功能单元。所谓的产品功能单元就是该款产品的基本规格。

步骤八：评估测算结果分析

目前，EPD的评估测算主要利用全生命期测算软件进行计算，输入数据来源部分是现场统计，部分是测算软件自带的数据库。数据库中的信息是基于国家和行业主管机构或社团发布的官方文件，并保证适时更新。由于地理位置、工艺水平、原材料的品质等不同，各国乃至于不同地区的数据库是有很大差异的。目前常用的测算软件有Simapro、Gibe等。本示例中的产品信息数据由制造商提供的，基础数据库的信息采用欧洲发布的有关行业数据（表4-10）。

通过上述测算结果，可以看出如下信息：

第一，在能源消耗和环境影响因素方面，使用阶段所占比重最大，其次是制造阶段，最后是废弃处理阶段；

第四章 国外绿色建筑产品

某企业生产的 PVC-U 窗 EPD 评估结果　　　　表 4-10

序号	项目	制造阶段	使用 10 年	使用 30 年	使用 50 年	废弃处理
1	不可再生能源（MJ eq）	1.3E+3	2.4E+3	7.1E+3	1.2E+4	1.7E+1
2	可再生能源（MJ eq）	5.7E+1	9.6E+0	2.8E+1	6.5E+1	2.5E−2
3	危险废弃物（kg）	2.1E−1	1.5E−3	1.1E−2	9.3E−2	/
4	其他废弃物（kg）	1.1E+2	1.0E+0	3.0E+0	1.3E+2	/
5	GWP（kg CO_2-eq）	7.5E+1	1.5E+2	4.4E+2	7.7E+2	1.2E+0
6	AP（kg SO_2-eq）	2.8E−1	1.8E−1	5.5E−1	1.1E+0	3.7E−3
7	ODP（kg R_{11}-eq）	2.8E−6	2.8E−5	8.4E−5	1.4E−4	1.6E−7
8	POCP（kg C_2H_4-eq）	2.9E−2	7.1E−2	2.1E−1	3.6E−1	9.8E−4
9	EP（kg PO_4-eq）	4.9E−2	3.6E−2	1.0E−1	1.9E−1	8.4E−4

第二，通过对 10 年、30 年和 50 年使用年限的平均估算，使用年限越长，相同使用期内总消耗越低。这里需要注意的是，10 年、30 年和 50 年的数据不具有直接比较性。如 50 年使用年限的 PVC-U 窗，生产和废弃处理环节只发生 1 次，对于 10 年使用年限的 PVC-U 窗，生产和废弃处理环节发生了 5 次。但这种测算忽略了一个因素，每 10 年更换一次窗，达到的性能（保温、采光等）应当比持续用 50 年的窗要好。

4. EPD 评估示例——卫生陶瓷

下面再以卫生陶瓷产品为例，说明不同产品在 EPD 评估的要点差异。

步骤一：确认产品评估技术依据

卫生陶瓷产品 EPD 评估依据包括：

1）ISO 14025：全生命期产品评价规则，主要规定了全生命期分析（LCA）、生命周期清单（LCI）和信息模块。企业对企业（BtoB）和企业对消费者（BtoC）两种不同的模式。

2）EN 15804：2013：建筑产品全生命期评价规则，主要规定了建筑产品 EPD 的可比性，用于建筑产品 PCR 的要求、EPD 证书内容要求。

3）PCR CPC54（版本 2012）：卫生陶瓷产品种类规则。

4）ISO 14040 和 ISO 14044：全生命期分析（LCA）方法。

步骤二：确认产品评估阶段

根据产品评估时包含的生命期不同阶段，EPD 分为三种类型（表 4-11、表 4-12）。

EPD 评价类型　　　　表 4-11

EPD 类型	包含的声明期阶段	单元	可比较性
Ⅰ 从摇篮大大门	仅且必须覆盖 A1-A3 阶段	宣称单元	不可比较
Ⅱ 从摇篮到大门（增加可选项）	必须覆盖 A1-A3 阶段，其余可选	宣称单元或功能单元	不可比较
Ⅲ 从摇篮到坟墓	必须覆盖 A1-C4 阶段，其余可选	功能单元	对于具有相同的功能单元可比较

第六节 环境产品声明 EPD

EPD 类型与生命周期关系 表 4-12

生命期阶段	产品			建造		使用阶段							废弃处理				系统边界外
类型	A1	A2	A3	A4	A5	B1	B2	B3	B4	B5	B6	B7	C1	C2	C3	C4	D
	原材料获取	运输至企业	制造	运输至工地	建造	使用	维护	维修	更换	翻新	运行阶段能源消耗	运行阶段水消耗	拆除	运输	废物处理过程	废弃处置	再利用可能
						使用年限设定											
EPD 类型 I	√	√	√														
EPD 类型 II	√	√	√	O	O	O	O	O	O	O	O	O	O	O	O	O	O
EPD 类型 III	√	√	√	√	√	√	√	√	√	√	√	√	√	√	√	√	O

注："√"必选；"O"可选。本例是针对 EPD 类型 I 进行的讨论。

步骤三：生产用原材料信息

该款卫生陶瓷产品各部分原材料的组成信息如表 4-13。

某生产企业卫生陶瓷原材料信息 表 4-13

序号	原材料	原材料比例（%）
1	黏土	25～35
2	长石	25～35
3	高岭土	20～30
4	砂子及其他辅料	10～20

步骤四：分析原材料对人体健康和环境的影响（表 4-14）

原材料对环境和健康的影响 表 4-14

序号	材料	环境影响	人体健康影响
1	黏土	/	/
2	长石	/	/
3	高岭土	/	/
4	砂子及其他辅料	/	/

对于上述原材料的影响进行评价的主要依据是《危险化学品目录（2015 版）》，这版是参照联合国《全球化学品统一分类和标签制度》(Globally Harmonized System of Classification and Labelling of Chemicals，简称 GHS）制定而成的。

步骤五：基于全生命期分析计算要点

对于建筑产品基于全生命期分析计算时应关注如下内容，表 4-15 是针对卫生陶瓷的示例。

第四章　国外绿色建筑产品

卫生陶瓷 EPD 统计计算要点　　表 4-15

序号	要点	说　明
1	功能单元/宣称单元	本例中功能单元是生产1吨的卫生陶瓷。由于卫生陶瓷产品种类繁多，平均每件按15公斤估算
2	目标和范围	本例中 EPD 考虑了1吨卫生陶瓷从摇篮到大门，附加废弃处置情况
3	系统边界	本例中包括如下环节： 原材料的获取和运输至生产工厂； 原材料的进一步加工和配套模具的生产； 产品全部生产过程，包括浇注、干燥、上釉、入窑等； 能源和水资源的消耗，废弃物管理； 成品发货前的包装
4	估计和假设	对于卫生陶瓷产品包装回收率依据（欧洲）有关法律不低于44%
5	取舍原则	全生命期分析考虑了所有原材料，未舍去任何部分
6	背景数据	原始数据的采集期是2013~2014年，包括原材料、电力、水资源、天然气和废弃物的统计。向大气的排放未进行直接测量，而是通过数据库中能源消耗进行换算。数据库中的基础数据为欧洲二手统计数据，根据具体生产厂所在地的情况采用了调整系数
7	数据质量	原始数据是从生产厂直接获得，通过 Simapro 软件进行测算
8	评估的期间	数据统计涉及的期间为2013~2014年
9	分配	全部生产过程未产生副产品，无须进行分配核算
10	比较	只有相同功能的产品依据 EN 15804 进行核算，才有可比性

步骤六：EPD 评估结果

基于上述计算要点，完成数据采集和处理后，得到如下结果：

1）生产1吨卫生陶瓷的资源使用情况见表4-16。

某企业卫生陶瓷资源使用情况　　表 4-16

参数	单位	制造阶段 A1-A3	废弃阶段 C4
可再生一次能源的使用（作为原材料使用的不计）	MJ	0.56	0
可再生一次能源的使用（作为原材料使用）	MJ	444	0
可再生一次能源的总用量	MJ	444.56	0
不可再生一次能源的使用（作为原材料使用的不计）	MJ	13470	268
不可再生一次能源的使用（作为原材料使用）	MJ	0	0
不可再生一次能源的总用量	MJ	13470	268
二次原料的使用	kg	0	0
可再生二次燃料的使用	MJ	0	0
不可再生二次燃料的使用	MJ	0	0
新鲜水的使用	M3	5.20	0

2）1吨卫生陶瓷输出流和废弃后的清单如下表4-17。

3）1吨卫生陶瓷对环境的影响见表4-18。

5. EPD 实施的可行性与意义

根据上述示例步骤，我国在绿色建筑产品领域开展 EPD 评价需要关注如表4-19。

第六节 环境产品声明 EPD

某企业卫生陶瓷输出和废弃信息　　　　　　　　　　　　表 4-17

参数	单位	制造阶段 A1—A3	废弃阶段 C4
有害的废弃物（HWD）	kg	0.246	/
无害的废弃物（NHWD）	kg	212	1000
有辐射的废弃物（RWD）	kg	/	/
可再用的部件（CRU）	kg	8.23E−02	2.86E−02
可循环用的材料（MFR）	kg	0.034	0.0494
可恢复能源的材料（MER）	kg	/	/
能源载体（EE）	MJ	/	/

某企业生产单位卫生陶瓷对环境的影响　　　　　　　　　　　表 4-18

序号	项目	单位	制造阶段	废弃处理
1	GWP	kg CCO_2-eq	855	20.8
2	AP	kg SCO_2-eq	2.78E+00	6.63E−02
3	ODP	kg CFC11-eq	6.52E−05	1.09E−06
4	POCP	kg C_2H_4-eq	1.48E−01	6.25E−03
5	EP	kg PO_4-eq	7.52E−01	5.07E−02

开展 EPD 评价的主要环节和资源需求　　　　　　　　　　　表 4-19

序号	项目	依据	备注
1	确认具体建材产品的 PCR	ISO 14025，PCR	国际上已发布了部分建材产品的 PCR
2	产品信息	制造商提供	需要第三方机构通过实验或现场检查核实
3	原材料组成分析	制造商提供，GHS	需要第三方机构通过现场检查核实
4	使用阶段能源消耗和环境影响	ISO 14040，ISO 14044	软件测算，需要分别建立我国不同能源消耗对环境影响因素、主要材料生产不同能源消耗、能源补偿采用设备对于环境影响因素的数据库
5	废弃处理阶段能源消耗和环境影响	ISO 14040，ISO 14044	软件测算，需要建立建筑垃圾分类回收利用率、回收处理用不同能源形式消耗量、回收处理不同材料的环境影响因素的数据库

通过上述示例，如下内容需要引起注意：

第一，对于部分建材产品而言，原材料的能源消耗和对环境因素的影响远大于产品的装配生产阶段，如门窗、暖通空调设备。

第二，确保产品品质，提高产品在建筑物中应具有的基本性能，对于降低能源消耗、减少对环境因素的不利影响才是最重要的。

第三，建筑垃圾分类处理能力有待进一步关注和提高。

第四，不同能源形式消耗而造成的环境影响因素数据有待建立完善。

第五，主要基础原材料生产所消耗的能源形式、数量以及原材料自身或化学反应造成的环境影响因素数据有待建立完善。

第六，PCR 的建立有待完善。

第四章 国外绿色建筑产品

第七，案例中使用阶段能源补偿计算合理性有待商讨，因为造成热交换不仅仅是PVC-U窗，还有围护结构等因素，但围护结构所占比例或权重应根据具体设计方案而异。

第八，相同产品用于不同气候地区和建筑设计时，在使用阶段最终测算结果会不同。

因此，在建筑领域推动EPD工作的目的不是简单评价建材产品的优劣，而是推动产业链上下游各环节的协同发展，实现绿色建筑产业的整体品质提升，落实我国"十三五"提出的生态文明建设。

第五章　我国绿色建筑产品评价

——天下难事必作于易，天下大事必作于细

本章主要阐述了如下内容：
- 产品、部品和建材的概念：鉴于政府不同文件中提到的建筑产品、部品和建材的名词，介绍了三个概念的相关性和差异性。
- 绿色产品的概念和特征：介绍了国内外对于绿色产品的主流定义；阐述了绿色产品有别于普通产品的重要特征。
- 绿色建筑产品评价体系：阐述了评价体系的构建原则、构建方法和体现宜具备的条件。
- 构建我国绿色建筑产品评价体系建议：分析了《绿色建材评价标识导则》的特征，验证了既有评价体系的水平，提出了进一步发展的建议。

第一节　产品、部品和建材的概念

近几年，我国政府对于绿色建筑产业的发展，对于生态文明建设非常重视。但是，在不同的政府文件中出现了不同的专有名词，主要是建筑材料、建筑产品和建筑部品，这三个名词的概念是有区别的，因此，首先明确绿色建筑产品的范畴是有必要的。

建筑材料是土木工程所用材料及制品的统称，是土木建筑的重要物质基础。建筑材料覆盖面广，种类繁多，可以按性质或性能进行分类。按性质可分为建筑用非金属材料、金属材料、有机材料和复合材料等。如建筑用非金属材料主要包括天然的黏土、砂砾、石材和人造的砖、瓦、陶瓷、琉璃等烧土制品，水泥、石灰、石膏等胶凝材料，以水泥为基础的各种混凝土、砂浆及其制品，各种玻璃及其制品，以及无机涂料、石棉、矿棉、纤维制品、熔岩制品、碳化制品等。建筑用金属材料主要包括以钢铁、有色金属及其合金制造的型材、管材、板材和金属制品等。建筑用有机材料主要包括木材、竹材、建筑塑料、有机涂料、胶结材料和有机类保温材料等。建筑用复合材料，从狭义上是指纤维增强塑料（比如玻璃钢）和层压材料；广义上则是指两种或两种以上的材料复合组成的材料，可包括很多人造建筑材料的品种，如各种水泥砂浆和混凝土被称作水泥基复合材料，还有沥青复合材料和钙塑制品等。建筑材料按用途可分为绝热材料、吸声材料、防水材料、灌浆材料、装饰材料等。

建筑产品是建筑工程从立项、设计、施工、使用到维修全过程中所用到的产品，包括各种硬件、流程性材料、软件、服务以及它们的组合。凡可供在建筑中单独使用的任何一种上述实体,就是一种建筑产品。

建筑部品是具有相对独立功能的建筑产品,是由建筑材料、单项产品构成的部件、构件的总成,是构成成套技术和建筑体系的基础。建筑部品是建筑物的主要组成部分,是建筑物的一个独立单元,它具有规定的功能,按照建筑物的各个部位和功能要求,建筑物进行部品的划分,使其在工厂内制作加工成半成品(即部件化),运至施工现场,达到在工程现场组装简捷、施工迅速的目的,并保证部品安装就位后能达到规定的技术要求和质量要求,发挥其功能作用。

通过上述概念描述可以看出建筑材料、建筑产品和建筑部品在指代范围上的侧重点及其区别。建筑产品的概念包含建筑材料,但暖通空调设备和集成化建筑产品未包含在建筑材料的定义中。建筑部品是建筑物的一个独立功能单元,在工厂内已制成半成品,在施工现场进行装配即可。因此,建筑部品也是建筑产品范畴的一部分。建筑产品的概念相较于建筑材料和建筑部品所包含的范围更加广泛。

第二节 绿色产品的概念和特征

绿色概念的出现是经济、社会和生活的各个方面都发生巨大变化的必然结果,这些变化主要体现在如下几个方面:

1. 认知的变化。绿色概念的出现改变了人们的传统思维观念,过去那种靠过度消耗自然资源以追求物质享受的生活方式遭到了否定。现实环境的恶化,以及对恶化原因的分析与宣传,从国家到个人都认识到问题的严重性,人类要世世代代幸福地生存,必须改变价值观和生活习惯,实现自然资源和生存环境的可持续发展。因此,从消费者、企业到政府部门的消费活动、生产经营计划的制定、产品的开发、政策法规的制订等方面都开始体现可持续发展思维。

2. 消费方式的变化。在人类社会的发展进程中,生产与消费相互作用。当生产发展到一定水平时,必然会引起消费方式、消费观念的变革,而消费方式、消费观念的变化又必然会促进生产和技术的进一步发展。然而,生产与消费的变化往往受社会经济环境的制约。绿色消费是指在保护广大消费者利益的前提下,为满足生态需要,购买和使用对环境有益的绿色产品的消费行为。绿色消费的内容非常宽泛,涵盖了生产行为、消费行为的方方面面。也就是说,在社会消费中,不仅要满足当代人的消费需求和安全、健康,还要满足子孙后代的消费需求和安全健康。

3. 生产方式的转变。随着经济全球化的发展,市场竞争日趋激烈。如何面对这种挑战,开发、生产具有市场竞争力的、高技术含量的绿色产品已成为制造业面临的严峻挑战。我国近些年来生产方式和理念也发生了重大的变化。从粗放式生产模式向精益化生产模式转变,从劳动密集型生产向技术密集型生产转变,从手工化向自动化乃至智能化转变,从制造大国向制造强国转型,应当说,云技术与物联网是工业4.0的基础，而工业 4.0 将颠覆我们对于生产方式的传统认知。工业 4.0 将在满足人民需求和个性化不断增长的前提下，实现绿色化工业生产。

绿色产品是 20 世纪 80 年代后期世界各国为适应全球环保战略，进行产业结构调整的产物。1972 年在瑞典的斯德哥尔摩召开的"联合国人类环境会议"首次提出了"人类只有一个地球"的环境宣言。1992 年在巴西的里约热内卢举行的"联合国环境与发展大会"通过了《环境发展宣言》和《21 世纪议程》的可持续发展战略已经开始在世界各国的社会、经济各相关方面着手实施。

绿色产品目前尚无确切的定义，不同学者从不同角度对绿色产品作了不同的定义，如：

1. 绿色产品是指生产、使用及处理过程中符合环保要求，对生态和环境无害、少害或有利于资源再生和回收利用的产品。

2. 绿色产品主要指产品的功能除满足产品的使用要求外，还必须从产品的设计、原材料选用、工艺技术、成品包装、市场流通、使用回收等全过程统筹考虑，最大限度地满足人们的使用需要，并切实做到保护环境，与环境相容。

3. 所谓绿色产品是指各种具备节能、易回收等减少环境污染特征的产品。

4. 绿色产品是能满足用户功能需求，并在其生命周期过程中（原材料制备、产品设计及制造、包装及发运、安装及使用维护、回收处理及再利用）能经济地实现节约资源和能源、减小或消除环境污染，并对生产者和使用者的健康具有良好保护的产品。

5. 1999 年在我国首届绿色建材发展与应用研讨会上提出绿色建材的定义。绿色建材是指采用清洁生产技术、不用或少用天然资源和能源，大量利用固态废弃物生产的无毒害、无污染、无放射性、达到使用周期后可回收利用，有利于环境保护和人类身体健康的建筑材料。

绿色产品区别于一般产品的重要特征主要表现在以下几个方面：

1. 产品的环境特性。绿色产品要在其整个生命周期中具有良好的环境特性，即从材料获取与加工、产品生产与制造、使用维护乃至淘汰废弃后的回收处理都应具有良好的环保特性，不危及人体健康与安全，不造成环境污染，正常生命周期内应具有良好的维护性，当其丧失使用功能时应具有良好的拆卸性和回收性，并具有一定的可重复利用性。

2. 产品的能源特性。任何产品在整个生命周期中必然涉及能源的输入和输出，所以，能源特征应是每一绿色产品不可缺少的重要特征要素。

3. 产品的功能特征。绿色产品首先应具有优良的规定的使用功能，同时应具有可靠的质量保证，具有较高的可靠性。

4. 产品的经济特征。绿色产品要力争在满足各种功能、使用性能等的前提下，降低产品成本。因此，产品的经济特征是产品各项特征的综合反映。绿色产品与传统产品的经济成本有显著区别，不仅包括其制造成本、使用成本、维护成本、能耗成本等，还应包括回收处理费用等。

第三节 绿色建筑产品评价体系

对绿色产品概念和特征的探讨分析，有助于建立绿色产品的评价指标体系。绿色产品的评价指标体系应该把产品质量、环境影响、资源消耗、能量消耗和经济性等作为重要的因素予以考虑。

产品质量指标主要是指产品在使用阶段有关性能的综合指标。

环境影响指标主要是指产品在生命周期全过程中可能会产生的对水环境、大气环境、土壤环境的影响等。

资源消耗指标是指所用资源（包括主要原材料和辅料）种类、资源特性、资源消耗状况，包括资源属于可再生资源还是不可再生资源，是稀有资源还是丰富的资源，资源的消耗量、利用率/损耗率、再生利用率等。

能源消耗指标是指能源种类、能源特性、能源消耗状况，包括所用能源属于一次能源或二次能源，短缺能源还是丰富能源等，能源的可利用或可再生特性，能源的消耗量、利用率、再生利用率等。

经济性指标主要是指产品全生命期的综合经济效益。它应是在产品的全生命期内采用绿色技术与不采用绿色改造技术所产生的成本之差，与采用绿色技术对应的治理措施与不采用绿色改造技术而达到同等效果对应的治理措施成本之差，两个成本差值的求和。

制定绿色产品评价指标体系的指导思想是：有利于生态环境保护，有利于资源和能源的有效利用，有利于人体的安全与健康，有利于提高产品的技术水平和市场竞争力。

绿色产品评价指标体系的制定必须遵循以下基本原则：

1. 综合性原则。指标体系应能全面反映评价对象的综合情况，应能从产品质量、资源消耗、能源消耗、环境影响和社会经济效益等方面进行评价，充分利用产品认证审查、产品性能检测、环境监测、EPD 和碳排放评价等合格评定方法，保证综合评价的全面性和可信度。

2. 科学性和先进性原则。在绿色产品评价指标体系中，其相关指标的意义应明确，力求客观、真实、准确地反映被评价对象的"绿色"属性。同时，指标的测定方法应标准，统计方法要规范。有些指标可能目前尚无法获取必要的数据，但与评价关系较大时仍可作为建议指标提出。同时，绿色产品的理念包涵现代设计技术和管理技术，指标体系应当有效地反映这些现代技术的基本特征。

3. 系统性原则。在综合指标的基础上，反映各个属性之间的协调性关系。协调性关系可以通过两种方式予以反映，一是通过指标权重予以反映，一是凝练出新的相对性评价指标，以反映综合指标之间的正相关、负相关或零相关性。

4. 动态指标和静态指标相结合。评价指标受政策导向、技术标准、用户需求等因素的制约，对绿色建筑产品的要求也将随着现代化工业技术，尤其是工业 4.0 的发展而不断变化。在评价中，既要考虑到现有状态，又要充分考虑到未来的发展空间。

第三节 绿色建筑产品评价体系

5. 定性指标与定量指标相结合。绿色建筑产品评价指标应尽可能量化，但某些指标量化难度较大，此时也可采用定性指标来描述，以便从质和量的角度，对产品对象作出科学的评价结论。

6. 可操作性原则。绿色建筑产品评价指标应以一定的行业统计数据作为基础，同时指标项目要适量可操作，所制定的指标，在不同的产品之间必须具有可比性。

7. 层次性原则。绿色产品评价指标体系为产品设计人员、管理部门及消费者提供了设计决策、产品检查及绿色产品消费选择的依据。由于使用对象不同，因此，应在不同层次采用不同指标。如对管理部门，需要知道的是产品设计总体指标对需求的满足程度，显然这一层次的指标着重于其整体性和综合性；而设计人员需要知道所选的具体方案满足特定要求或功能的程度，这时的指标应更细致、更明确。因此，在不同层次上应有不同的指标。同时，产品生命周期是有层次的，指标体系应考虑其层次性。

8. 产品的差异性原则。不同行业的企业，其外部环境、企业内部层次等情况差异很大，所以，产品在设计、制造过程中，对环境的影响也有差异，对所有的产品采用统一的评价指标是不妥当的。但过于细致地划分产品，又会增加评价的难度，故应将功能相近的产品归为一类，适当地制定评价指标。

参照上述这些原则，并结合建筑产品特点，理想的绿色建筑产品评价指标体系宜具备三个条件：

1. 指标之间具有可比性，即采用统一的原则和标准选取指标。
2. 指标表达形式简单化，对指标进行简化处理同时保持最大信息量。
3. 指标之间具有联系性，需要进行指标产生机理研究，将指标统一在一个综合框架中。

指标体系的基本结构一般有四种形式，即多指标体系、树型指标体系、丛型指标体系和矩阵指标体系。

多指标体系是最简单的指标体系结构，该结构的特点是从多个特征指标上即可直观地对系统进行评价，指标元素除同属一个集合外，无其他关系，且相互独立。该结构的评价指标体系优点在于简洁、直观；不足之处在于不适合用于复杂的评价系统。

树型指标体系的结构是一个没有回路的连通图。树的结点集合就是指标元素不同层级的多重子集。

丛型指标体系的结构与树型结构的共同点是，评价指标体系指标间具有层次性的隶属关系。不同之处在于，树型结构的下级指标只唯一隶属于它的上级指标；而丛型结构的下级指标可以隶属于两个或以上的上级指标。

矩阵指标体系的结构实际上是一种两维表结构，通常可用有限矩阵来表示。

由于建筑产品的功能不同、制造工艺方法不同、所用材料不同等众多因素的影响，建筑产品的绿色特性往往有不同的表现。同时，随着新技术、新工艺和新材料等的应用，绿色产品的评价指标本身也在不断地变化。因此，绿色建筑产品评价指标体系要能够尽可能地涵盖产品绿色特性的各个方面并具有通用性，可以说，指标体系的构建是有难度和挑战性的。

构建评价指标体系的方法主要有调研法、目标分解法和多元统计法等。调研法是通过直接或间接调查研究方法收集相关评价指标，利用比较分析方法进行归类分析，并根据评价目标设计评价指标体系，以问卷的形式把所设计的评价指标体系发给有关领域的专家进行审阅，通过对反馈意见的汇总分析，对其进行最终确定。该方法的优点在于获取信息比较全面，评价指标覆盖面积较广；不足之处在于由于选取指标过多可能会造成指标重复或繁琐。

目标分解法是通过对评价目的进行具体逐层分解来构建评价指标体系。从对评价目的开始逐次分解，直到分解出来的指标达到可测的要求。该方法的优点在于构建的评价指标体系具有清晰的层次结构，指标间的隶属关系简单明了；不足之处在于选择评价指标的形式可能过于单一，不够全面。

多元统计法是通过聚类分析、主成分分析和因子分析等统计方法，从初步拟定的评价指标进行聚类分析，找出拟定指标之间的有机联系，然后再利用主成分分析法找出初拟定的指标体系中起关键作用的指标，最后对初拟定指标进行因子分析，指出评价指标体系中各指标的主次位置。该方法的不足之处在于需要具体的指标数据，计算过程较为复杂。

第四节　构建我国绿色建筑产品评价体系建议

2015年10月住房城乡建设部、工业和信息化部联合印发了《绿色建材评价技术导则（试行）》（第一版）的技术文件。在该《导则》文件中，明确给出了绿色建材的定义是指在全生命周期内可减少对天然资源消耗和减轻对生态环境影响，具有"节能、减排、安全、便利和可循环"特征的建材产品。应当说，这个定义从全生命期角度出发，考虑了与"绿色"相关的主要特点，评价内容明确，对于绿色建材的评价工作具有指导性的意义。《导则》中评价指标体系构成如表5-1。

《导则》评价指标体系绿色特征分析　　　　表5-1

评价性质	评价内容	评价方式	对应的绿色特征
控制项	大气污染物排放	评审监测报告	环境指标：生产阶段
	污水排放	评审监测报告	环境指标：生产阶段
	噪声排放	评审监测报告	环境指标：生产阶段
	工作场所环境	评审监测报告	环境指标：生产阶段
	安全生产	评审安全生产标准化证书	环境指标：生产阶段
	管理体系	评审体系认证证书	环境指标：生产阶段；产品指标：生产阶段
	基本性能	评审产品检测报告	产品指标：使用阶段
	应用技术文件	评审企业提供的技术文件	产品指标：生产阶段、使用阶段
	其他：个性化产品指标，比如燃烧性、放射性等	评审产品检测报告	产品指标：使用阶段

第四节 构建我国绿色建筑产品评价体系建议

续表

评价性质	评价内容	评价方式	对应的绿色特征
评分项			
节能	1. 单位产品生产能耗或碳排放（通用）	评审企业统计报表	能源指标：生产阶段
节能	2. 原材料运输能耗（通用）	评审企业统计报表	能源指标：原材料获取（运输）阶段
节能	3. 能源管理体系认证（通用）	评审体系认证证书	能源指标
节能	4.1 单位产品淡水消耗（砌体材料）	评审企业统计报表	资源指标：生产阶段
节能	4.2 导热系数（保温材料）	评审检测报告	产品指标：使用阶段
节能	4.3 强度等级（混凝土）	评审检测报告	产品指标：使用阶段
节能	4.4 热工性能（节能玻璃）	评审检测报告	产品指标：使用阶段
节能	4.5 陶瓷砖厚度	评审检测报告	产品指标：使用阶段
节能	4.6 单件重量、用水效率、洗净功能（卫生陶瓷）	评审检测报告	产品指标：使用阶段
减排	1. 厂区大气污染物和污水排放	评审监测报告	环境指标：生产阶段。比控制项要求等级高。
减排	2. 产品认证、EPD、碳足迹（通用）	评审认证证书、评估证明	产品指标（产品认证）：使用阶段；环境指标（EPD、碳足迹-部分指标）：全生命期；能源指标（EPD、碳足迹）：全生命期；资源指标（EPD）：全生命期
减排	3.1 不使用氟氯烃发泡剂和六溴环十二烷阻燃剂（保温材料）	评审企业生产工艺文件和记录	环境指标：生产阶段
减排	3.2 清洁生产水平（节能玻璃）	评审与 HJ/T 361 的等级符合度	能源指标：生产阶段；环境指标：生产阶段；资源指标：生产阶段
减排	3.3 放射性污染（陶瓷砖、卫生陶瓷）	评审检测报告	产品指标：使用阶段
减排	3.4 冲水噪声（卫生陶瓷）	评审检测报告	产品指标：使用阶段
减排	3.5 大气污染物（不含颗粒物）排放、颗粒物排放、普通砂浆散装率和特种砂浆袋装率（预拌砂浆）	评审监测报告；评审生产包装技术有关材料	环境指标：生产阶段；产品指标：生产阶段
安全	1. 安全生产标准化水平（通用）	评审安全生产标准化证书	环境指标：生产阶段
安全	2.1 干燥收缩率、吸水率（砌体）	评审检测报告	产品指标：使用阶段
安全	2.2 抗冻性（砌体）	评审检测报告	产品指标：使用阶段
安全	2.3 抗压强度、块体密度（砌体）	评审检测报告	产品指标：使用阶段
安全	2.4 燃烧性能（保温材料）	评审检测报告	产品指标：使用阶段
安全	2.5 结构连接安全性（保温材料）	评审检测报告或其他说明材料	产品指标：使用阶段

第五章 我国绿色建筑产品评价

续表

评价性质	评价内容	评价方式	对应的绿色特征
安全	2.6 标准差（混凝土）	评审企业检测报告	产品指标：生产阶段
	2.7 抗渗等级、抗氯离子渗透等级、抗碳化等级、抗冻等级（混凝土）	评审检测报告	产品指标：使用阶段
	2.8 施工安全性能（节能玻璃）	评审企业提供的证明材料	产品指标：使用阶段
	2.9 可见光反射比（节能玻璃）	评审检测报告	产品指标：使用阶段
	2.10 使用安全性能（陶瓷砖）	评审检测报告	产品指标：使用阶段
	2.11 强度（预拌砂浆）	评审检测报告	产品指标：使用阶段
	2.12 强度离散系数（预拌砂浆）	评审企业提供的证明材料	产品指标：生产阶段
	2.13 耐久性能（预拌砂浆）	评审检测报告	产品指标：使用阶段
	2.14 测量管理体系认证（预拌砂浆）	评审认证证书	产品指标：生产阶段
便利	1. 适应性和经济性（通用）	专家评分	经济指标
	2. 施工性（通用）	专家评分	产品指标：使用阶段
	3.1 尺寸精度（砌体）	与标准相比	产品指标：使用阶段
	3.2 施工过程环境影响（保温材料）	专家评分	环境指标：使用阶段
	3.3 施工性能、自密实混凝土（混凝土）	与标准相比	产品指标：使用阶段
	3.4 一般显色指数（节能玻璃）	评审检测报告	产品指标：使用阶段
	3.5 单件包装重量（陶瓷砖）	评审企业提交的材料	产品指标：生产阶段
	3.6 建筑模数要求（陶瓷砖）	评审企业提交的材料	产品指标：使用阶段
	3.7 烧成后无须后加工（陶瓷砖）	评审企业提交的材料	产品指标：生产阶段
	3.8 耐污染性（陶瓷砖）	评审企业提交的材料与标准相比	产品指标：使用阶段
	3.9 安装、更换和维护（卫生陶瓷）	专家评分	产品指标：使用阶段
可循环	回收和再利用	根据企业提交的材料，专家评分	资源指标：生产阶段
	废弃物利用	根据企业提交的材料，专家评分	资源指标：生产阶段
	报废混凝土产生率（混凝土）	根据企业提交的材料，专家评分	资源指标：生产阶段
	工业废水回收利用比例（混凝土）	根据企业提交的材料，专家评分	资源指标：生产阶段
	低质原料使用量（陶瓷砖、卫生陶瓷）	根据企业提交的材料，专家评分	资源指标：生产阶段
	灰料利用（预拌砂浆）	根据企业提交的材料，专家评分	资源指标：生产阶段
加分项	采用先进工艺和设备，且环境影响低于行业平均水平	根据企业提交的材料，专家评分	环境指标：生产阶段
	材料性能有创新性且优于行业平均水平	根据企业提交的材料，专家评分	产品指标：使用阶段

第四节　构建我国绿色建筑产品评价体系建议

根据对于上述《导则》的梳理分析，我们从全生命期角度对评价指标体系的覆盖度进行验证。

第一，环境指标（表5-2）。

全生命期环境影响因素　　　　　　　　　表5-2

环境影响因素	原材料获取（含运输）	产品生产过程	产品使用过程	产品废弃回收过程
大气环境	OK	OK	OK	OK
水环境	OK	OK	OK	OK
土环境	OK	OK	OK	OK

第二，资源指标（表5-3）。

全生命期资源影响因素　　　　　　　　　表5-3

资源影响因素	原材料获取（含运输）	产品生产过程	产品使用过程	产品废弃回收过程
可再生（定性绝对）	OK	/	/	/
不可再生（定性绝对）	OK	/	/	/
丰富（定性相对）	OK	/	/	/
稀缺（定性相对）	OK	/	/	/
消耗量（定量绝对指标）	OK	OK	/	/
利用率（定量相对指标）	OK	OK	/	/
可回收率（定量相对指标）	OK	OK	/	OK

第三，能源指标（表5-4）。

全生命期能源影响因素　　　　　　　　　表5-4

能源影响因素	原材料获取（含运输）	产品生产过程	产品使用过程	产品废弃回收过程
丰富	OK	OK	OK	OK
稀缺	OK	OK	OK	OK
一次能源（可再生/不可再生）	OK	OK	OK	OK
二次能源	OK	OK	OK	OK
消耗量（定量绝对指标）	OK	OK	OK	OK
利用率（定量相对指标）	OK	OK	OK	OK
可回收率（定量相对指标）	OK	OK	OK	OK

第四，指标汇总比对情况（表5-5）。

全生命期指标体系覆盖情况汇总　　　　　　表5-5

影响因素	原材料获取（含运输）	产品生产过程	产品使用过程	产品废弃回收过程
产品性能	/	已涉及	已涉及	/
资源	OK	已涉及（节能玻璃）	/	/
能源	已涉及运输	已涉及	OK	OK
环境	OK	已涉及	已涉及	OK
经济	OK	OK	OK	OK

通过评价指标比对验证，得出如下主要信息：

1)《导则》的评价指标体系在我国现有建材行业统计数据的基础上，从全生命期角度比较全面地覆盖了建筑产品的主要绿色特征性指标。

2) 建筑产品生产过程中对于原材料的需求，尤其是考虑到原材料作为资源的属性和利用率，还需进一步研究完善；

3) 建筑产品的原材料获取过程对于能源的需求信息现状比较匮乏。

4) 由于目前《导则》中的建材产品都不属于用能产品，所以产品使用过程中对于能源的需求指标还未体现。

5) 产品废弃回收阶段对于能源的需求统计数据比较薄弱。

6) 建筑产品所需原材料在获取阶段以及产品废弃回收阶段对于环境的影响，所能应用的评价技术还有待进一步开发。

7) 建筑产品在全生命期对于经济性指标的评价技术还需进一步开发。

8) 生产阶段能源消耗数量已经纳入评价范围，不同能源的形式在现阶段评价文件中尚未予以关注，存在进一步识别的空间。

9) 绿色建筑产品信息量化指标需要进一步提高，综合运用大数据和云计算技术为建筑设计选型服务。

因此，充分认识绿色建筑产品概念和评价指标体系对于政策研究具有重要的指导作用。

第六章 认证制度与政策、法律和法规

——法者,天下之准绳也

本章主要阐述了如下内容:
- 社会管理理论:简要阐述了该理论发展的主要历史阶段,重点阐述了社会管理的职能和方式,讲述了第三方机构参与社会管理的理论依据。
- 新公共管理理论:介绍了公共产品和公共服务,阐述了"掌舵人"和"划桨人"分开的理论,重点介绍了社会组织和第三方机构在该理论下应发挥的作用。
- 政策执行过程模式:简要介绍了政策执行过程模式、互动理论模式和综合模式。
- 我国绿色建筑产品认证的政策环境:依据政策执行模式理论,结合社会管理理论的方式和新公共管理理论的第三方作用,从政策自身完善程度、政策执行机构执行力、目标群体接受程度、政策环境适宜性等四个维度分析绿色建筑产品认证政策落实绩效。

第一节 社会管理理论

正如在开篇介绍合格评定的发展历程一样,产品认证不是人类社会与生俱来的产物,而是社会发展到一定阶段,为了满足社会的需求而出现的。产品认证作为一种政府职能的演变与延伸,与社会管理理论体系的发展,尤其是公共产品理论的发展对于政府职能的要求密切相关。

我们可以给社会管理作一个概念性的描述,即社会管理就是社会的管理主体(如政府和/或社会组织等)运用多种资源和方式,为实现客体(人)最大限度的自由和发展,为了促进社会系统协调运转,对客体(人)形成的社会系统的各个组成部分的不同领域以及社会发展的各个环节,进行规范、组织、协调、控制、监督和服务的全过程,其目的是为了满足社会成员生存和发展的基本需求,解决社会问题,提高人民生活质量,其本质是以人为中心的管理和服务。

社会管理既是弥补"市场失灵"的必然要求,也是协调各种矛盾与冲突的必要前提。在一定意义上讲,社会管理主要是以强制性行政手段为基础,以法律为保障,对社会关系进行调整和约束,政府在其中起着非常重要的主导作用。对于涉及社会整体的公共利益,需要依靠国家与政府权力加以解决。政府管理的社会事务,其主旨在于通过实施有效的社会政策,达到保障公民权利、维护社会秩序、协调社会利益、管理

第六章 认证制度与政策、法律和法规

社会组织、提供社会安全和解决社会危机等目的。因此,制定政策和法规是政府干预社会的主要手段和基本措施。

根据绝大多数西方学者对于政府社会管理理论的研究文献,以社会政策作为基本判断标准,可以将政府社会管理发展历程大致划分为三个历史发展阶段,即自由竞争市场经济时期的政府有限社会管理职能阶段、混合经济时期的政府全面社会管理职能阶段或称福利国家社会政策阶段、全球化市场经济时期的政府社会政策阶段。

在自由竞争市场经济时期,政府奉行不干预政策,对社会发展基本上采取放任自由的态度,社会管理主要以社会的自我管理和社会自治为主。政府维护社会秩序的主要目标,在于保护财产权,进而维护其建立在财产权基础上的社会秩序,而一些非政府组织在此时起到了补充救济的职能。

在混合经济时期,社会保障成为政府社会管理的工作重点。政府以促进充分就业、维护宏观经济稳定作为施政的重要目标。在这个时期,政府的社会管理同时起到弥补市场失灵与社会失灵的双重作用。政府主要致力于建立和完善基本社会关系管理制度、建立资本与劳动合作的社会制度、完善社会主要利益集团围绕政府与公共支出的多数表决制度,同时,政府也致力于发展社会自治和社会自我管理。

在全球化市场经济时期,针对混合经济时期社会保障出现的"消极福利国家"的弊端,社会管理理论着力研究如何将"消极福利国家"转变为"积极福利国家"、"工作福利国家"或"社会投资型国家"。社会政策的重点和主题是围绕着国家、社会、社区、家庭、个人在福利中的地位和作用展开。相关学派的倡导者主张要彻底改革福利国家制度,变消极的福利制度为积极的福利制度,以充分就业政策为核心,将"福利"转变为"工作",并适度限制福利支出的增长,达到平衡经济发展与社会保障发展、需求管理与供给管理、社会管理与经济增长的目标。

社会管理是一个有丰富含义而又论说不一的复杂概念,而且社会管理概念随着各国的文化传统、政治体制、经济社会发展状况的不同也带有各自不同的特点。主要表现在如下方面:第一,发达国家的社会管理主要指政府力量对独立于政治、经济领域之外的那部分社会公共事务的管理。一方面,这种管理提供国家必需的基本秩序;另一方面,现代政府社会管理的主要方向是为经济发展和人民生活提供一个稳定良好的社会环境。第二,发达国家的社会管理相比较而言更具有"规制"的色彩。在发达国家整体"服务型政府"的背景下构成了其刚性但又不可或缺的职能。这里所说的社会管理职能主要由政府力量来行使,但是在欧美国家,越来越多的私营部门和第三方机构在获得许可后已介入部分社会管理过程。

从这个概念出发,我们可以概括出社会管理有以下的职能和方式:

社会管理的基本职能主要有以下四个方面:第一,社会控制。这是社会管理的首要职能。通过运用法律、行政手段和道德及舆论的力量,对社会成员与群体的越轨行为进行谴责、制止和制裁,以维护正常的社会秩序。第二,社会保障。在现代社会保障制度建立之前,社会保障最早是对某些社会成员的救济,是对穷人的一种恩赐。接受救济者往往以牺牲人格尊严、人身自由、政治权利等为代价。进入现代社会之后,

社会保障从一种恩赐变为一种权利。其义务主体除了政府之外，还有受法律、行政手段强制而必须对特殊人群履行法定义务的社会成员与群体。第三，社会服务。这主要指社会服务的主体应积极发展社会公共事业，进行公共设施建设，向全社会提供公共产品和公共服务。第四，社会协调。随着社会的发展，社会问题出现综合化、复杂化的趋向，仅靠某一群体、某一方面很难有效解决问题，需要通过社会管理主体，加强社会协调组织各方面力量共同解决，实现社会全面进步。

社会管理一般而言，主要包括以下几种方式：第一，统筹规划与政策引导。主要是指通过制定社会发展的规划和计划，明确社会发展长期和近期目标、战略任务等。通过政策，向全社会鲜明地表明鼓励什么、怎么鼓励，禁止什么、如何禁止等，端正社会发展方向，协调社会关系，解决社会矛盾和社会问题，推动社会全面进步，实现社会的健康协调发展。第二，由法律法规加以约束。通过立法和执法活动，规范人们的日常行为，打击制裁危害社会违法犯罪行为，协调社会关系，保护社会成员合法权益，维护社会正常秩序。第三，调节经济利益。主要指通过财政拨款、资助和信贷、税收等手段，调动全体社会成员与群体投身社会发展，积极参与社会公益活动。第四，行政控制。通过行政命令、行政措施及其行政活动，分配社会资源，组织社会活动，推动社会事业发展。第五，舆论宣传与思想道德教育。向社会成员宣传国家法律法规和各项方针政府及行政命令，提高其自我约束能力，增强其法制意识，提高其道德水准和思想觉悟，从而提高整个社会的文明程度。

第二节 新公共管理理论

"新公共管理"理论力图通过在公共管理过程中引入市场竞争机制，实现不断提高公共管理水平及公共服务质量的目的。该理论的基本思想有：第一，掌舵与划桨分开。把服务与执行职能从掌管它们的集中决策部门分离出去，给予服务提供和执行机构更大的灵活性和自主性，并通过与这些机构签订绩效合同使其对服务结果负责。第二，政府广泛采用私营部门成功的管理手段经验和竞争机制。引入竞争机制，取消公共服务供给的垄断性，让更多的私营部门参与公共服务的供给，通过这种方式提高服务供给的质量和效率。创新社会管理体制的核心是建立新型的政府与社会的关系。西方国家的社会管理模式，有一个共同点，就是它们健全完善社会覆盖面极广的社会组织和第三方机构，这些机构在社会管理中充当了政府与社会公众沟通的桥梁和媒介，为缓和社会矛盾起到了缓冲器的作用。社会管理涉及的领域宽，关系到不同阶层、群体的多元化利益，仅靠政府难以承担如此繁重复杂的社会公共事务，因此，政府要积极构建复合社会管理模式，通过职能让渡、政策激励、制度改革促进第三方机构承担部分社会服务职能，具体而言，是承担政府所转移的部分职能。社会经济快速发展，要求政府转变工作职能，将其本不应承担的社会职能剥离，由社会组织承接，集中精力抓好对社会发展重大政策和战略的研究、制定及实施，加强宏观层面的管理，提升决策的质量，由"划桨者"变为"掌航人"，由"全能政府"变为"有限政府"，提高社会管理工作的效率，形成"小政府、大社会"的格局。

新公共管理的对象是社会公共事务。但不同学科对公共事务的理解侧重点不同，目前学术界对于公共事务的内涵尚未取得共识。哲学和社会学理解的公共事务是指个人在公共活动中的体验。经济学理解的公共事务是指社会所需要的公共物品，这种物品具有效用扩展于他人的成本为零，但也无法排除他人享用的属性。公共管理所研究的公共事务主要是指一定时期与一定共同体成员共同利益相关的社会事务，体现为一定共同体成员普遍需求的公共产品。

任何公共事务的背后都体现和反映着一定的公共利益，因此，维护、分配和增进一定共同体的公共利益成为公共管理的宗旨。公共管理主体必须以公众的公共需求为导向，以公共需求的实际满足为衡量标准。随着公共事务治理主体多元化与运作机制的弹性化，善治成为实现和增进公共利益的途径。善治的本质特征是政府与社会利益相关方对公共产品的合作管理，其衡量标准包括合法性、透明性、责任性、有效性、参与性、稳定性、公正性等方面。

新公共管理实现渠道主要包括以下几个方面：第一，行政授权。公共管理提倡建立小而优的政府，这样势必把相当一部分的职能和职权转移给社会第三方机构，这种转移主要以行政授权的方式实现。第二，行政指导和行政合同。随着政府放松管制，强制性的消极行政权逐渐退却，而一些协商性的积极行政行为粉墨登场，日益成为公共管理的主流方式而被广泛采用。因此，新公共管理的具体方式就是发挥第三方机构的作用。政府可以在不同地区、不同领域，通过采用与第三方机构签约、公私合作等形式开展不同程度的实践工作和示范项目。

第三节 政策执行过程模式

政策制定的难点，也是重点之一，是政策的执行落地情况。由于政策在执行过程发生重大偏离而导致政策的初衷无法落实的情况不乏先例。因此，在了解了社会管理理论和新公共管理理论后，我们很清楚，对于公共产品可以通过政策和法规，引入第三方执行机构推动相关社会职能工作，但为更好地实现政策执行的效果，也为了识别政策执行发生偏离的风险，这里介绍政策执行过程模式理论。政策执行过程模式有三个主要模式，即政策执行过程模式、互动理论模式和综合模式。

政策执行过程模式是由史密斯提出的。该模式认为，政策执行过程所涉及的因素主要包括理想化的政策、目标群体、执行机构和环境因素。

互动理论模式是由麦克拉夫林提出的。该模式指出政策执行过程就是政策的执行组织和政策的受影响者之间就目标手段作出的相互调适的互动过程。

综合模式是由萨巴蒂尔和马泽曼尼安提出的。该模式认为影响政策执行各个阶段的因素，最主要的可以分成三大类：①政策问题的可办性，主要包括现行有效的理论与技术、目标团体行为的多样性、目标团体的人数、目标团体行为需要调适的幅度；②政策本身的规制能力，主要包括政策本身含有充足的因果论、明确的政策指令、充分的财政资源、执行机构间与机构内部的层级整合、建构执行机关的决定规则、征募执行的人员、安排外界人士参与的机会；③政策本身以外的变数，主要包括经济环境与技术、媒介对问题注意的持续、大众的支持、赞助团体的态度与资源、监督机关的支持、员工的热忱与领导技术。

影响政策执行的具体因素在现实中各有不同，但总的来说，一个理想政策的执行模式必须不断地与环境各因素之间建立联系。影响政策执行的因素基本可以涵盖在以下四个维度中：第一，政策自身完善程度指标，具体包括政策自身目标的合理性、公平性、连续性，自身的技术难度，政策的明确性，以及政策稳定性；第二，政策执行机构执行力指标，具体包括执行机构的人力财力资源、信息资源、权威资源，机构本身内部结构的合理性、制度资源，内部人员的素质，以及执行人员对政策的认知、认可度等；第三，目标群体接受程度指标，具体包括目标群体对政策认同程度、心理变化、成本—收益情况以及配套政策的针对性；第四，政策环境适宜性指标，具体包括社会政治环境、经济环境、制度环境、文化环境、资源能源环境等。

第四节 我国绿色建筑产品认证的政策环境

绿色建筑产品从其属性而言，具有公共产品的特征。我国政府结合国家发展战略，提出生态文明建设总体方案，确保政策的有效实施，提高人民福祉和幸福指数。我们依据政策执行模式理论，从下面四个维度阐述绿色建筑产品认证在落实过程中的重要事项。

1. 政策自身完善程度

2013年1月国家发展改革委员会、住房城乡建设部发布《绿色建筑行动方案》，明确提出研究建立绿色建材认证制度，编制绿色建材产品目录，引导规范市场消费。2015年10月住房城乡建设部、工业和信息化部联合发布《绿色建材评价标识管理办法实施细则》和《绿色建材评价技术导则（试行）》，标志着绿色建材评价进入实施阶段。2015年9月中共中央国务院印发《生态文明体制改革总体方案》，明确提出建立统一的绿色产品体系，将目前分头设立的环保、节能、节水、循环、低碳、再生、有机等产品统一整合为绿色产品，建立统一的绿色产品标准、认证、标识等体系。完善对绿色产品研发生产、运输配送、购买使用的财税金融支持和政府采购等政策。

表6-1汇总了目前部分部委分头设立的有关节能、低碳等认证与标识制度情况。

有关部委节能、低碳、绿色产业指导意见一览　　　　　　　　表6-1

序号	部委	时间	文件	表述	备注
1	发改委 质检总局	2004.8	能源效率标识管理办法	用能产品能源效率等级等性能指标的一种信息标识，属于产品符合性标志的范畴。	节能

第六章 认证制度与政策、法律和法规

续表

序号	部委	时间	文件	表述	备注
2	国务院办公厅	2007.7	关于建立政府强制采购节能产品制度的通知	节能效果显著、性能比较成熟的产品	节能
3	环保总局	2001.12	环境保护产品认定管理办法	指用于防治环境污染、改善生态环境、保护自然资源的设备、环境监测专用仪器和相关的药剂、材料	环保,该文件已作废
4	环境保护部	2008.9	中国环境标志使用管理办法	在生产、使用及处置等过程中采取一定措施消除污染或减少污染,达到中国环境标志产品技术要求,并通过中国环境标志认证的产品	环保
5	发改委 认监委	2013.2	低碳产品认证管理暂行办法	指与同类产品或者相同功能的产品相比,碳排放量值符合相关低碳产品评价标准或者技术规范要求的产品	低碳,该文件已作废
6	质检总局 发改委	2015.11	节能低碳产品认证管理办法	节能产品认证是指由认证机构证明用能产品在能源利用效率方面符合相应国家标准、行业标准或者认证技术规范要求的合格评定活动;低碳产品认证是指认证机构证明产品碳排放量值或者温室气体排放量符合相应低碳产品评价标准或者技术规范要求的合格评定活动	节能; 低碳
7	住建部 工信部	2014.5	绿色建材标识评价管理办法	指在全生命周期内可减少对天然资源消耗和减轻对生态环境影响,具有"节能、减排、安全、便利和可循环"特征的建材产品	绿色
8	中共中央 国务院	2015.9	生态文明体制改革总体方案	将目前分头设立的环保、节能、节水、循环、低碳、再生、有机等产品统一整合为绿色产品	绿色

通过上述文件可以看出,对于绿色产品从我国政策上具有稳定性和延续性。相关部委都在所辖范围内根据党中央国务院的指导精神开展了有关工作。从绿色建筑产品角度而言,目前政策引导和舆论宣传工作已经开展,但在技术法规标准、调节经济利益和行政控制方面还有待进一步完善。

2. 政策执行机构执行力

政府在选择第三方机构承担社会公共服务或提供公共产品时,执行机构对于政策的认知和认可度,以及执行机构相关的资源,包括制度资源、权威资源、人力资源、信息资源等直接关系到政策实施的成败。因此,政策执行机构的选择是政策执行过程中重要的一环。政府在利用社会第三方资源时,对执行机构的期望值应是规划好的,是明确的。

作为社会第三方政策执行机构,首先要清楚政策的目的和意义,以及机构自身在政策中承担的责任和起到的作用,必须清楚认识到机构行为对于社会的影响,从而在思想上对承担的政府工作高度重视。

政策执行机构在承担工作前，应对政策有高度的认可。执行机构自身的认可度越低，信心就越弱，执行力就不强，会影响政策的落实贯彻程度。因此，执行机构相关人员执行工作前思想认识的高度统一有利于政策的落实。

执行机构的相关资源反映了机构自身能力条件。如何从纷杂的社会资源中选择合适的组织机构承担相应的政府分派的职能工作，需要考虑执行工作的性质和既有制度的相容性，从而决定对执行机构特点的要求。以目前中央的指导精神和有关部委的文件，采用合格评定制度推动绿色产品的落实工作。之所以采用合格评定制度，首先是因为合格评定制度已纳入我国的法制体系，有法可依；其次，合格评定制度是国际公认的质量控制和性能评价制度，有助于我国在国际合作项目以及区域性合作工作中接轨；最后，我国已将认证认可纳入了高新技术服务产业，2016年认证认可从业机构已被国家统计局依据《中华人民共和国统计法》纳入了行业统计范围。

选择认证机构承担绿色建筑产品认证执行工作，应对认证机构在该领域的专业能力和人力资源的配套能力有所要求。我国目前已获批准的认证机构逾200家，其中可以从事产品认证的机构逾100家，有能力从事绿色建筑产品认证的机构从专业能力而言数量有限。执行机构专业能力不足，削弱政策技术权威性，从而影响目标群体对于政策态度的严肃性。

随着进入大数据云平台时代，政策执行机构如何实现与政府委托部门、社会相关利益方有关信息的共享也是考量的内容之一。相关信息资源共享有助于解决信息不对称而产生的问题，有助于政策执行过程的公平性和透明性，有利于接受政府和社会相关利益方的监督。

3. 目标群体接受程度

政策能否得到有效落实，同样取决于政策目标群体对政策的认同程度。绿色建筑产品的目标群体基本可以分为三级，分别为直接群体、间接群体和终端群体。直接群体是绿色建筑产品的生产企业和制造商，间接群体是绿色建筑产品的有关应用单位，如设计单位、施工单位和开发商，终端群体是建筑的最终使用者。通过近些年的宣传和消费理念的变化，终端消费群体是认可绿色概念的，但目前从市场角度而言，终端群体对建筑在建造阶段是否使用绿色建筑产品没有话语权。间接群体在招采阶段是有话语权和决定权的。直接群体虽然有话语权，而且与绿色建筑产品密切相关，但缺乏决定权。

因此，了解目标群体和目标群体的话语权及决定权之后，可以分析政策所产生的成本收益关系对目标群体的影响，以及目标群体会有什么样的反应。现阶段，推行绿色建筑产品，对于直接目标群体制造商而言，从原材料选取、生产工艺、技术创新等方面需要进行必要的技术投入和设备改造，会增加制造成本。对于间接群体而言，一旦采用绿色建筑产品，每平方米造价会有所提高。但由于使用绿色建筑产品而造成的材料成本提高的比例占总造价比例值有限。

因此，采用配套的政策方式对绿色建筑产品实施管理，发挥着至关重要的作用。通常可以采用的政策方式有三种：强制性措施、引导性措施、激励性措施。强制性措施是通过行政规定发布实施；引导性措施可以通过标准等技术手段和项目招采阶段的要求进行规制；激励性措施通过财政补贴或金融优惠政策得以实现。

对于绿色建筑产品采用的是引导性和激励性措施适用于我国建筑市场的现状。对于绿色建筑产品制定必要的技术标准是落实绿色建筑产品政策的技术基础，可以起到引导性作用。激励措施宜根据谁投入、谁受益的原则，鼓励直接技术的投入方，即制造商（包括生产厂）获益。而且，通过财政激励，可以降低生产成本，避免多米诺骨牌效应。

4. 政策环境适宜性

党的十八大报告明确提出"加强和创新社会管理，推动社会主义和谐社会建设"的目标任务，加强和创新社会管理，提高社会管理科学水平，是社会建设的重要内容。

十八大在深入分析社会组织发展现状和研究社会管理规律的基础上，明确提出建立健全"五位一体"，即党委领导、政府负责、社会协同、公众参与、法治保障的管理模式。

（1）党委领导，是指要充分发挥党委在社会管理中总揽全局和协调各方的领导核心作用，组织动员基层党组织和广大党员积极投身于直接为人民群众服务的工作。

（2）政府负责，是指政府应提供更多更好的社会公共服务，使政府及各职能部门的管理更加协调有效，确保由政府负责的社会管理和公共服务人员到位、投入到位、工作到位、责任到位，完善法规政策，健全社会管理体系，培育发展和管理监督好社会组织，畅通公民参与渠道等，切实发挥政府在社会管理中的主导作用。

（3）社会协同，是指工青妇等群众组织、基层群众性自治组织、社会组织、企事业单位要发挥协同作用，形成党委和政府与社会力量互联、互补、互动的社会管理和公共服务网络，把大量社会性、公益性、事务性管理起来，充分发挥协同、自治、自律、他律、互律作用。

（4）公众参与，是人民群众依法依规，理性有序参与社会管理和公共服务，实现自我管理、自我服务、自我约束、自我发展。

（5）法治保障。一是目标层面，就是指依照宪法和法律规定管理社会事务，保证各项管理工作都依法进行，逐步实现社会主义民主的制度化、规范化、程序化。根本目标是为了解决管理理念问题，强调树立社会主义法治理念、运用法治思维，用"宪法法律至上"原则来指导各项社会管理创新工作；二是在操作层面，强调用法治的方式应对矛盾和解决问题，就是进一步发挥法治在社会管理中的作用，提出新思路和新要求，推进法治意识、法治机制、法治氛围在实践层面得以尽快落实，通过完善各项法律制度，努力改进法律实施的途径和方法，提高依法办事的能力，转变立法、执法和司法等法律实施工作的态度和作风，将现代管理学的理念，特别是整体性治理的工作思路引入法治建设领域，增强基于法治原则管理社会的效率，建立完善行政职能的法律赋权与法律监督机制，形成公共治理新格局，从根本上维护社会稳定和繁荣，全面建设和谐的社会秩序。

深化行政体制改革的重要内容是"创新行政管理方式，提高政府公信力和执行力，推进政府绩效管理",要求政府职能向创造良好发展环境、提供优质公共服务、维护社会公平正义转变。党的十八届三中全会印发了《中共中央关于全面深化改革若干重大问题的决定》,对从当前至 2020 年我国重要领域和关键环节的改革作出了重要部署,以"促进社会公平正义、增进人民福祉"作为新一轮改革的出发点和落脚点。随着经济社会的发展、行政环境的变化以及建设人民满意的服务型政府的推进,公众对政府提供的公共产品和公共服务提出了更高的要求。公共价值作为一种全新的公共管理理念,反映的是政府的行为逻辑,强调的是公共产品的公共效用与政府在行动过程中对公众行为的公益导向。建设服务型政府是新时期社会管理的必然要求。政府要充分践行经济调节、市场监管、社会管理、公共服务四项职能。

通过树立多方参与、共同治理的理念,充分发挥社会组织在社会管理中的协同作用。在我国社会主义和谐社会的构建过程中,政府在逐步由原来的"管理型"政府向"服务型"政府转变,在这一转变过程中,必然会让渡一部分社会管理职能给社会第三方机构,于是,社会组织便形成了企业、个人和政府之间的桥梁。社会第三方机构相当于国家与社会之间的一个中介力量,它既能够实现群体利益的下情上达,又有助于国家政策在群体中的贯彻落实。

第七章　认证制度与标准体系

——工欲善其事，必先利其器

本章主要阐述了如下内容：
- 现有标准体系框架简介：简要介绍了标准层次和标准级别，说明了基础标准、通用标准和专用标准、国家标准、行业标准、地方标准、社团标准和企业标准的特征和侧重点。
- 绿色建筑产品标准体系探讨：基于既有的国际和国外全生命期评价标准体系，结合我国的现状，提出了绿色建筑产品标准体系建议。
- 绿色建筑产品与绿色建筑：结合第四章的介绍，提出了以 EPD 为核心评价技术，GreenFormat 为标准信息表达方式，BEES 决策支持系统为主体的产品与建筑的结合方式。

第一节　现有标准体系框架简介

标准可以根据专业技术层次和发布设立级别进行分类。我国目前的标准层次通常可以分为基础标准、通用标准和专用标准。标准级别包括国家标准、行业标准、地方标准、社团标准和企业标准。

基础标准通常在一定范围内既可以直接应用，又可以作为其他标准的依据和基础，具有普遍的指导意义。基础标准是在某领域中覆盖面最大的标准。它几乎是该领域中其他标准的共同基础。

基础标准主要包括以下几类：

1. 技术通则类：如"电子工业技术标准制修订工作有关规定和要求"、"设计文件编制规则"等。这些技术工作和标准化工作规定是需要全行业共同遵守的。

2. 通用技术语言类：如制图规则、术语、符号、代号、代码等。这类标准的作用是使技术语言达到统一、准确和简化。

3. 结构要素和互换互连类：如公差配合、表面质量要求、标准尺寸、螺纹、齿轮模数、标准锥度、接口标准等。这类标准对保证零部件互换性和产品间的互联互通、简化品种、改善加工性能等都具有重要作用。

4. 参数系列类：如优先数系、尺寸配合系列、产品参数、系列型谱等。这类标准对于合理确定产品品种规格，做到以最少品种满足多方面需要，以及规划产品发展方向，加强各类产品尺寸参数间的协调等具有重要作用。

5. 环境适应性、可靠性、安全性类：这类标准对保证产品适应性和工作寿命以及人身和设备安全具有重要作用。

6. 通用方法类：如试验、分析、抽样、统计、计算、测定等各种方法标准。这类标准对各有关方法的优化、严密化和统一化等具有重要作用。

通用标准是将具有相同特征归纳到一起，制定一个在一定技术领域内皆可使用的标准。以工程技术领域为例，许多个性标准之间往往包含一些共性的特征，如产品标准中的尺寸规格、参数系列、使用环境条件、验收规则和试验方法等。

专用标准是对于一类具体产品制定的性能评价判定的标准。如产品标准可以视为专用标准。

根据《中华人民共和国标准化法》的规定，对需要在全国范围内统一的技术要求，应当制定国家标准。国家标准由国务院标准化行政主管部门制定。对没有国家标准而又需要在全国某个行业范围内统一的技术要求，可以制定行业标准。行业标准由国务院有关行政主管部门制定，并报国务院标准化行政主管部门备案，在公布国家标准之后，该项行业标准即行废止。对没有国家标准和行业标准而又需要在省、自治区、直辖市范围内统一的工业产品的安全、卫生要求，可以制定地方标准。地方标准由省、自治区、直辖市标准化行政主管部门制定，并报国务院标准化行政主管部门和国务院有关行政主管部门备案，在公布国家标准或者行业标准之后，该项地方标准即行废止。企业生产的产品没有国家标准和行业标准的，应当制定企业标准，作为组织生产的依据。企业的产品标准须报当地政府标准化行政主管部门和有关行政主管部门备案。已有国家标准或者行业标准的，国家鼓励企业制定严于国家标准或者行业标准的企业标准，在企业内部适用。

第二节 绿色建筑产品标准体系探讨

绿色建筑产品是经历了一个发展过程而形成的，这其中有一些概念和配套标准的发展起了至关重要的作用。

节能标准的发展，确切地讲，是建筑节能指标的提高推动了建筑产品的升级和创新。建筑节能推动建筑产品有关指标的提高主要体现在具体的产品标准中。

以环保部推动的"中国环境标识"认证，即"十环认证"是通过发布行业标准（HJ），产品标准中仅关注与环保有关的性能指标作为认证的技术依据。所以，无论节能还是环保产品，我国的标准都是针对具体产品中与节能或环保有关的具体指标进行修订。

随着全生命期概念的出现，碳足迹标准的发布，温室气体全生命期统计等一系列国际和国外标准的实施，尤其是环境产品声明（EPD）概念的推出，对于建筑产品全生命期"绿色"程度的评价标准体系产生了重大的影响。

表 7-1 是基于全生命期评价所涉及的国际和国外标准。

第七章 认证制度与标准体系

表 7-1 国际和国外主要环境标准一览

序号	标准号	标准名称	备注
1	ISO 14001—14009	环境管理体系系列标准	
2	ISO 14010—14019	环境审核系列标准	
3	ISO 14020—14029	环境标志系列标准	ISO 14021：自我申明（Ⅱ型） ISO 14024：产品认证（Ⅰ型） ISO 14025：产品全生命期认证（Ⅲ型）
4	ISO 14030—14039	环境表现评价系列标准	
5	ISO 14040—14049	生命周期评价系列标准	ISO 14044：环境管理-产品生命周期评价-要求和导则
6	ISO 14050—14059	术语和定义	
7	ISO 14060	产品标准中的环境指标	
8	ISO 14064—1	组织层次上对温室气体排放和清除的量化和报告的规范及指南	
9	ISO 14064—2	项目层次上对温室气体减排和清除增加的量化、监测和报告的规范及指南	
10	ISO 14064—3	温室气体声明审定与核查的规范及指南	
11	ISO 14065	温室气体——对从事温室气体合格性鉴定或其他形式认可的确认与验证机构的要求	
12	PAS 2050	商品和服务在生命周期内的温室气体排放评价规范	
13	EN 15804	建筑工程可持续发展-环境产品声明-建筑产品类别核心规则	
14	PCR	一类具体的产品类别规则	

其中，ISO 14000 系列标准的框架如图 7-1 所示。

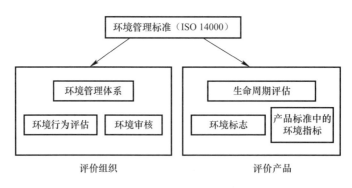

图 7-1 环境标准分类框架

ISO 14020 系列标准规定了第三方认证评价方法（ISO 14024，Ⅰ型认证）、企业自我声明方法（ISO 14021，Ⅱ型认证）以及产品全生命期认证评价评价方法（ISO 14025，Ⅲ型认证）。以上标准是从评价层面给出的规范性要求。EN 15804 是针对建筑产品全生命期评价而给出的规范性要求。PCR 是针对具体一类产品给出的评价方法。上述标准虽然给出了评价方法，但并未给出不同产品在性能方法和"绿色"程度方面具体的判定指标。产品性能方面的判定是依据既有产品标准。"绿色"程度是基于全生命期统计和估算给出的具体的数值，但未对数值高低进行优劣评判。

综合上述标准的梳理和归类分析情况，建议我国绿色建筑产品认证标准体系参考 ISO 14020 系列、ISO 14040 系列和 ISO 14060 系列为基础标准，参考 EN 15804 作为通用标准，以 PCR 为参考，结合现有《绿色建材评价技术导则》作为专用标准。

第三节 绿色建筑产品与绿色建筑

建筑产品不同于一般日用消费品，需要充分考虑和建筑应用的结合。因此，对于绿色建筑产品而言，前期与设计的有效结合，废弃后与回收再利用工艺的结合是绿色建筑产品有别于普通建筑产品的特点。

目前，国外的绿色建筑评价已将绿色建筑产品的使用率纳入考量范围。如英国可持续住宅标准节材评价体系中，重点关注的指标可以分为材料的环保性、材料的集中采购、垃圾处理及施工现场控制四方面。材料指标的关注重点集中于材料对环境的影响和建筑材料的集中采购，通过这两方面标准的提升实现建筑节材性能。在 LEED-NC 中节材评价指标主要分为生活垃圾处理与回收、建筑垃圾的处理与回收、建筑材料再利用、建筑材料优选优化四大类。其中，节材评价体系主要包括再生物存放和收集、建筑再利用、建设废弃物管理、材料再利用、循环材含量、地方材、快速再生材、认证的木材等评价指标。此外，BEES 决策支持系统的推出对于绿色建筑产品选型发挥了重要的作用。德国与英国和美国不同，德国作为生态建筑和被动式建筑技术领先的欧洲国家，自工业革命以来就拥有了一套完善的高要求的工业节能标准，而这套工业节能标准适合德国自身的国情，所以德国并没有刻意地在早期就推出一套完整且系统的绿色建筑评价体系。但随之 BNB 和 DGNB 中都提出了对于建筑产品的导向要求。

我国以《绿色建筑评价标准》GB/T 5038 为代表，近些年推出了从工业建筑到校园、医院等一系列绿色建筑标准。随着绿色建筑产品认证和标识体系的发展完善，绿色建筑产品目录的形成，必将助力绿色建筑的健康发展和评价标准的完善。

因此，建立绿色建筑产品数据库，为建筑设计和产品选型服务，是绿色建筑产品与绿色建筑有效结合的渠道。在这方面，美国的 GreenFormat 产品数据信息格式和 BEES 决策支持系统为我们提供了很好的借鉴。结合第四章国外已开展的相关工作的介绍和我国发展的实际情况，我国的绿色建筑产品应当从评级向性能和经济的量化方向发展，以 EPD 为核心评价技术，结合 BIM 等设计中产品选型需求，制定与 GreenFormat 相当的标准信息表达规范，同时增加数据电子化接口格式要求，应用相当于 BEES 决策支持系统。尤其是随着"十三五"我国装配式建筑的发展，建筑产品和预制构配件和部品的信息标准化，以及与建筑设计接口的标准化需要加快完善，而且，由政府统一组织构建绿色建筑产品信息平台有助于规范化管理和配套政策的实施，从而推动供给侧改革和企业品牌建设，拉动产业发展。

第八章 认证制度与政府管理

——道有因有循，有革有化

本章主要阐述了如下内容：
- 风险概念：介绍了风险的基本特征，基于社会管理等理论分析了政府征信第三方机构的潜在风险。
- 委托—代理理论：介绍了该理论的概念和形成，阐述了该理论的基本理论模式和前提条件。
- 委托—代理模型的理论解决方案：介绍了"劣币逐良币"等现象的理论依据，针对理论模式分析了风险分担和监督激励机制。
- 构建绿色产品认证制度探讨：结合政策执行模式和委托—代理理论解决方案，基于我国绿色建筑产品认证和评价的发展，提出了逆向选择和声誉激励等机制的建议。

第一节 风险概念

当政府将建筑产品作为准公共产品纳入管理范围，并拟采用采购社会服务方式进行管理时，正如政策执行模式所阐述的，需要解决执行机构的选择问题。选择标准的制定和实施过程也是风险识别和风险控制的过程。根据美国学者威雷特（1901年）对风险的定义，"风险是关于不愿发生的事件发生的不确定性之客观体现"。这一定义强调了两点：第一，风险是客观存在的，是不以人的意志为转移的；第二，风险的本质是"不确定"。任何事物都有其自身的特征和发展规律，认识风险所具有的特征，对于风险管理具有重大的意义，风险一般具有如下特征：

1. 客观性。人类存在的历史证明，无论是自然界的物质运动，还是社会发展规律，都由事物的内部因素所决定，由超越于人类主观意识而存在的客观规律决定的，因而对于风险管理而言，所能做的是在一定范围内改变风险形成和发展的条件，降低风险发生的概率，减少风险带来的损失。

2. 偶然性。风险虽然客观存在，但就某一具体风险而言，风险的发生具有偶然性，或者说是随机性。在发生之前，人们无法准确预测风险何时会发生以及发生的后果。这是因为导致任何一个具体风险的发生必须是诸多因素共同作用的结果，而每一个因素的作用时间、作用点、作用方向、顺序、作用强度等都必须是在满足一定条件的情况下，才能导致风险的发生。而每一个因素的出现本身就是偶然的。风险发生的偶然性还意味着在时间上具有突发性，在后果上往往具有灾难性。

3. 可预测性。我们可以根据以往发生过的类似事件的统计资料，通过概率分析，对某种风险发生的频率及其造成损失的程度作出主观上的判断和经验上的总结，从而对可能发生的风险进行预测和衡量。风险分析的过程实际上就是风险预测和衡量的过程。

4. 潜在性。风险不是显现在表面的东西，是在一定的条件下转变为现实的，具有潜在性。潜在性是风险存在的基本形式，认识风险的潜在特征，对于预防风险，具有重要意义。

5. 可变性。风险的可变性是指风险在一定条件下可转化的特性。任何事物都处于运动和变化之中，这些变化必然会引起风险的变化。风险的可变性包括风险性质和风险量的变化，也包括某些风险在一定的空间和时间范围内可被消除，但同时产生新的风险等。

对于第三方机构承担政府所管理的部分社会职能时，可能由于第三方机构的专业技术能力、机构自身利益、人员道德等问题，导致政策的执行落实与要求存在偏差甚至背离，从而产生负面社会影响。为了识别和管理由此可能产生的潜在风险，应用委托—代理理论可以有效控制此类风险的发生。

第二节　委托—代理理论

委托—代理理论就其实质来说，是一种契约理论，其基本内容阐述了某一当事人（委托人）聘用另一当事人（代理人）完成某项工作，而代理人又比委托人拥有更多的有关此项工作的信息（存在信息不对称），委托人可能无法观测到代理人努力工作的水平，而代理人也知道委托人无法确定其努力工作的程度（这一点是委托人和代理人之间的公共知识），此时代理人就有可能为了使自身效用达到最大化而采取机会主义行为（如偷懒、规避风险等），进而损害到委托人的利益的问题，即委托—代理问题。

委托—代理理论的形成，是从寻求系统地解决代理问题的方法开始的，这主要归功于威尔森（1969年）、斯宾塞和泽克梅森（1971年）、罗斯（Ross，1973年）、莫里斯（1974年，1976年）、霍姆斯特姆（1979年）、格罗斯曼和哈特（1983年）。他们开创了三个模型来分析和解决委托—代理问题，这三个模型是："状态空间模型化方法"、"分布函数的参数化方法"、"一般化分布方法"。这三个模型以递进的方式来表达委托人应采取何种行动，从而使自身效用达到最大化。三个模型共同的假设是：委托人对有随机的产出没有直接的贡献，代理人的行为不易直接地被委托人观察到。因而得出两个基本观点：在任何满足代理人参与约束和激励相容约束而使委托人预期效用最大化的激励合同中，代理人都必须承担风险；如果代理人是风险中性者，那么可以通过使代理人承担完全风险的合约以达到最优结果。这三个模型推导出的解决方案的出现，标志着委托—代理理论的产生。

第八章　认证制度与政府管理

构成委托—代理关系必须具备以下三个条件：第一，市场中存在两个相互独立的个体，双方都在约束条件下追求自身效用的极大化；第二，双方都面临市场的不确定性和风险，双方所掌握的信息处于不对称状态，代理人在交易中掌握更多的信息，有信息优势，委托人掌握的信息少，处于信息劣势；第三，代理人的行为影响委托人的利益，即委托人不得不为代理人的行为承担风险。

委托—代理理论就是关于非对称信息下交易的学说，其研究在不完备合约的情况下，如何更有效地激励代理人，并对代理人的行为进行监督和约束。非对称信息指的是某些参与人拥有另一些参与人不拥有的信息。信息的非对称可以分为两类：非对称发生的时间和非对称信息的内容。从非对称发生的时间区分，非对称性可能发生在当事人签约之前（exante），也可能发生在签约之后（expost），分别称为事前非对称和事后非对称。研究事前非对称信息博弈的模型称为逆向选择模型，研究事后非对称信息的模型称为道德风险模型。从非对称信息的内容区分，非对称信息可能是指某些人参与的行动，也可能是指某些人参与的知识，研究不可观测行动的模型称为隐藏行动模型，研究不可观测知识的模型称为隐藏知识模型或隐藏信息模型（表8-1）。

委托—代理理论基本模型一览　　　　　　　　　　表8-1

非对称性要素	要素划分	理论模型
时间	签约前	逆向选择模型
	签约后	道德风险模型
内容	行为	隐藏行为模型
	知识	隐藏知识/信息模型

对于上述描述的四种基本理论模型，根据实际情况，可以组合如（表8-2）。

委托—代理中的主要问题　　　　　　　　　　表8-2

要素		时间	
		签约后	签约前
内容	知识	隐藏信息的道德风险	信号传递模型
	行为	隐藏行动的道德风险	信号甄别模型

隐藏信息的道德风险模型描述的是签约时双方信息是对称的（因而是完全信息），签约后，代理人根据观测到的实际市场信息，然后选择行动（如向委托人报告实际信息），委托人观测到代理人的行动，但不能观测到实际的信息（因而是不完全信息）的情况。

隐藏行动的道德风险模型，签约时信息是对称的（因而是完全信息），签约后，代理人选择行动（如工作是努力还是不努力），代理人的行动和市场情况一起决定某些可观测到的结果，委托人只能观测到结果，而不能观测到代理人的行动本身和市场信息本身（因而是不完全信息）。

逆向选择模型，代理人知道自己的类型，委托人不知道（因而信息是不完全的）。典型的逆向选择模型有信号传递模型和信号甄别模型。

信号传递模型描述代理人知道自己的类型，委托人不知道（因而信息是不完全的），为了显示自己的类型，代理人选择某种信号，委托人在观测到信号之后与代理人签订合同。

信息甄别模型描述代理人知道自己的类型，委托人不知道（因而信息是不完全的），委托人可以提供多个合同供代理人选择，代理人根据自己的类型选择一个最适合自己的合同，并根据合同选择行动。

第三节 理论解决方案

从委托—代理理论来看，就是明确代理人的职责，理清委托人与代理人之间的权利和义务关系，制定合理的风险分担机制和有效的监督激励机制。

解决委托—代理问题的关键就在于建立一套完善的激励和监督体制，从而实现风险分担的目的。通常情况下，委托人只能观测到代理人的行为结果，即我们通常所说的绩效。因此，针对代理人的道德风险，委托人的问题就是设计一个激励和监督机制以诱使代理人从自身利益出发选择对委托人最有利的行动。在信息对称情况下，代理人的行为是可以被观察到的，委托人可以根据观察到的代理人行为对其实行奖惩。此时，帕累托最优风险分担和帕累托最优努力水平都可以达到。在非对称信息情况下，委托人不能观测到代理人的行为，只能观测到相关变量，这些变量由代理人的行动和其他外生的随机因素共同决定，于是委托人的问题是选择满足代理人参与约束和激励相容约束的激励合同以最大化自己的期望效用。

市场上著名的逆向选择就是"劣币逐良币"现象。因此，逆向选择会干预市场的有效运行，从而导致市场交易的低效率或无效率。为了避免出现逆向选择的结果，市场参与者都试图明确产品或服务价格与其质量之间的关系，就是"优质优价"问题。这种关系学术上用"市场信号"的概念来描述。通过有效的信息收集，委托人和代理人之间通过信号传递和信号甄别可以有效解决逆向选择问题。如高水平的机构可利用信息优势向委托人（政府）传播机构的私有信息，以证明机构的资质和专业实力，可以采用信号传递手段。信号传递和信号甄别的差别主要在于，在信号传递中，拥有私人信息的一方先行动，而在信息甄别中，不拥有私人信息的一方先行动。委托人可通过制定一套策略或契约来获取代理人的信息。这就是项目实施前标书的招采阶段。

一旦招采阶段结束，确立了委托—代理关系，进一步需要解决的是道德风险问题。在解决道德风险问题上，通常有三种方式，即合同激励、声誉激励和委托人监督。

合同激励是指代理机构提供的是高新技术服务（如认证或评价），由于信息不对称，委托人（政府）不能观测到代理机构（评价机构）的行动选择（是否客观、公正、认真等），只能观测到最终的代理结果（评价的数量和发证的数量）。委托人（政府）可通过合同激励代理人选择委托人希望的行动。

如果委托—代理关系不是一次性的而是多次性的，或者委托方与代理方存在长期

第八章　认证制度与政府管理

合作关系，即使没有显性激励合同，"时间"本身也可能会解决代理问题。当代理人的行为很难甚至无法证实，显性激励机制很难实施时，长期的委托代理关系就有很大的优势，长期关系可以利用"声誉激励"实现。拉德纳（Radner）和鲁宾斯坦（Rubbinstein）使用重复博弈模型证明，如果委托人和代理人保持长期的关系，帕累托一阶最优风险分担和激励是可以实现的。也就是说，在长期的关系中，第一，由于大数定理，委托人可以相对准确地从观测到的变量中推断代理机构的努力水平。第二，委托人对代理机构的资信水平、管理能力有相对更准确的了解，可以有效防止代理机构的"偷懒"或者与相关方"串谋"等损害委托人利益的行为。另外，为了能获取后续项目的代理权，代理机构会十分注重自己在市场上的声誉，因而会努力工作，不断提高自身能力水平。而且，如果代理机构在某个项目上发生"偷懒"、"串谋"等行为，即使可以获得超额的收益，但其违约机会成本太大，付出的将是失去长期合作的代价，因而代理机构也会选择努力工作。

委托人和代理机构的关系实际上是处于一种信息不对称的状态，由于评价项目的专业性和复杂性，委托人无法在事前对代理机构行为进行预测，同时由于代理机构业绩指标的可计量程度低等原因，委托人也无法在事后对代理机构的行为后果进行准确的评价和控制。这样委托人在与代理机构签订代理契约时就必须考虑如何减弱契约关系的信息不对称和提高代理人的行为可观察性，委托人不但要激励代理机构，还要对代理机构加以监督，并通过适当的机制提高代理机构的努力工作程度。

第四节　构建绿色产品认证制度探讨

建设工程通常工程周期长，隐蔽工程多，专业技术性强。工程质量问题有施工原因，也不乏产品质量原因。目前工程监管的管理制度以入场复检为主要手段，通过抽样复检可以实现对抽检批次复检项目性能的判定，但对于全部入场产品质量的稳定性依然缺乏总体判断。对于绿色建筑产品中"绿色"程度更不是通过抽检检测的项目能够验证的。因此，绿色建筑产品认证从性能和"绿色"程度的评价信息是对入场复检的有益而必要的补充。正如在开篇所介绍的，产品认证与检测不同，产品认证涵盖的范围要比检测报告大。此外，产品认证通过对生产厂质量控制能力的评审，尤其是对产品生产一致性和可追溯性的评审，能够更大程度地确保产品生产的稳定性。

当政府实施简政放权，使更多的社会组织，尤其是第三方认证机构参与社会公共产品和公共服务的管理活动中，通过委托第三方认证机构实施产品认证工作并采信其结果时，需要了解和控制委托机构的能力和执行力。结合上面委托—代理理论的阐述，我们不难发现，政府委托认证机构代理其行使产品认证活动时，事前风险即对于认证机构能力的了解，可以通过逆向选择模型解决，这就是我们通常采用的招标过程要求的一系列资质证明手段。这个阶段的核心问题是制定选择执行机构的条件。通常考虑的是三方面条件：第一，合法合规的资质条件；第二，技术能力、行业权威性和背景；

第三，拟授权机构与主管部门常年合作的关系。第二和第三方面的条件已经在政策执行模式和委托—代理理论声誉激励章节予以阐述。

众所周知，招采阶段后，在执行阶段，委托—代理风险依然存在，即道德风险。认证机构会因为利益最大化的驱动而发生隐藏信息和/或隐藏行动的道德风险问题。因此，采取必要的激励和监督机制是控制风险的有效手段。鉴于我国国情，以及调研阶段各有关部委开展的相关工作，可以看出，我国目前主要采取的是监督机制。这种监督机制包括主动监督（如例行检查或飞行检查）或被动监督（比如对于投诉和举报的调查处理）。此外，建议政府通过对于委托认证机构的绩效评价，在建筑领域采取合同激励和声誉激励模式。

随着大数据时代的到来和云技术的应用，代理方（认证机构）诚信平台的建立可以大幅降低信息不对称性，通过数据和图像的实时上传可以降低代理方在执行过程中隐藏信息或隐藏行动的风险。

附录 建筑用基础材料碳排放数据库示例

INVENTORY OF CARBON & ENERGY (ICE) SUMMARY

Materials	Embodied Energy & Carbon Coefficients			Comments
	EE-MJ/kg	EC-kg CO_2/kg	EC-kg CO_2e/kg	EE= Embodied Energy, EC= Embodied Carbon
Aggregate				
General (Gravel or Crushed Rock)	0.083	0.0048	0.0052	Estimated from measured UK industrial fuel consumption data
Aluminium	Main data source: International Aluminium Institute (IAI) LCA studies (www.world-aluminium.org)			
General	155	8.24	9.16	Assumed (UK) ratio of 25.6% extrusions, 55.7% Rolled & 18.7% castings. Worldwide average recycled content of 33%.
Virgin	218	11.46	12.79	
Recycled	29.0	1.69	1.81	
Cast Products	159	8.28	9.22	Worldwide average recycled content of 33%.
Virgin	226	11.70	13.10	
Recycled	25.0	1.35	1.45	
Extruded	154	8.16	9.08	Worldwide average recycled content of 33%.
Virgin	214	11.20	12.50	
Recycled	34.0	1.98	2.12	
Rolled	155	8.26	9.18	Worldwide average recycled content of 33%.
Virgin	217	11.50	12.80	
Recycled	28	1.67	1.79	
Asphalt				
Asphalt, 4% (bitumen) binder content (by mass)	2.86	0.059	0.066	1.68 MJ/kg Feedstock Energy (Included). Modelled from the bitumen binder content. The fuel consumption of asphalt mixing operations was taken from the Mineral Products Association (MPA). It represents typical UK industrial data. Feedstock energy is from the bitumen content.

附录 建筑用基础材料碳排放数据库示例

续表

Materials	Embodied Energy & Carbon Coefficients			Comments
Asphalt, 5% binder content	3.39	0.064	0.071	2.10 MJ/kg Feedstock Energy (Included). Comments from 4% mix also apply.
Asphalt, 6% binder content	3.93	0.068	0.076	2.52 MJ/kg Feedstock Energy (Included). Comments from 4% mix also apply.
Asphalt, 7% binder content	4.46	0.072	0.081	2.94 MJ/kg Feedstock Energy (Included). Comments from 4% mix also apply.
Asphalt, 8% binder content	5.00	0.076	0.086	3.36 MJ/kg Feedstock Energy (Included). Comments from 4% mix also apply.
Bitumen				
General	51	0.38-0.43 (?)	0.43-0.55 (?)	42 MJ/kg Feedstock Energy (Included). Feedstock assumed to be typical energy content of Bitumen. Carbon dioxide emissions are particularly difficult to estimate, range given.
Brass				
General	44.00	2.46 (?)	2.64 (?)	Poor data availability. It is believed that the data may be largely dependent upon ore grade. Poor carbon data, making estimate of embodied carbon difficult.
Virgin	80.00	4.47 (?)	4.80 (?)	
Recycled	20.00	1.12 (?)	1.20 (?)	
Bricks				
General (Common Brick)	3.00	0.23	0.24	
EXAMPLE: Single Brick	6.9 MJ per brick	0.53kg CO_2 per brick	0.55	Assuming 2.3kg per brick.
Limestone	0.85	?	—	
Bronze				
General	69.0 (?)	3.73 (?)	4.0 (?)	Average of the only two references
Carpet				
General Carpet	74 (187 per sqm)	3.9 (9.8 per sqm)	—	For per square meter estimates see material profile. Difficult to estimate, taken from Ref. 94.
Felt (Hair and Jute) Underlay	19.00	0.97	—	Ref. 94.

附录 建筑用基础材料碳排放数据库示例

续表

Materials	Embodied Energy & Carbon Coefficients			Comments
Nylon (Polyamide), pile weight 300g/m2	130 MJ per sqm	6.7 (GWP) per sqm	6.7 (GWP) per sqm	Total weight of this carpet 1,477g/m². See Refs. 277 & 279. These carpets (inc. below) are a tufted surface pile made of 100% nylon (polyamide) with a woven textile backing and flame proofed on the basis of aluminium hydroxide.
Nylon (Polyamide), pile weight 500g/m2	180 MJ per sqm	9.7 (GWP) per sqm	9.7 (GWP) per sqm	Total weight of this carpet 1,837g/m2. See Refs. 277 & 279.
Nylon (Polyamide), pile weight 700g/m2	230 MJ per sqm	12.7 (GWP) per sqm	12.7 (GWP) per sqm	Total weight of this carpet 2,147g/m2. See Refs. 277 & 279.
Nylon (Polyamide), pile weight 900g/m2	277 MJ per sqm	15.6 (GWP) per sqm	15.6 (GWP) per sqm	Total weight of this carpet 2,427g/m2. See Refs. 277 & 279.
Nylon (Polyamide), pile weight 1100g/m2	327 MJ per sqm	18.4 (GWP) per sqm	18.4 (GWP) per sqm	Total weight of this carpet 2,677g/m2. See Refs. 277 & 279.
Carpet tiles, nylon (Polyamide), pile weight 300g/m2	178 MJ per sqm	7.75 (GWP) per sqm	7.75 (GWP) per sqm	Total weight of this carpet 4,123g/m2. See Refs. 277 & 279. These carpet tiles (inc. below) are a tufted surface pile made of 100% nylon (polyamide) fleece-covered bitumen backing and flame-proofed on the basis of aluminium hydroxide
Carpet tiles, nylon (Polyamide), pile weight 500g/m2	229 MJ per sqm	10.7 (GWP) per sqm	10.7 (GWP) per sqm	Total weight of this carpet 4,373g/m2. See Refs. 277 & 279.
Carpet tiles, nylon (Polyamide), pile weight 700g/m2	279 MJ per sqm	13.7 (GWP) per sqm	13.7 (GWP) per sqm	Total weight of this carpet 4,623g/m2. See Refs. 277 & 279.
Carpet tiles, nylon (Polyamide), pile weight 900g/m2	328 MJ per sqm	16.7 (GWP) per sqm	16.7 (GWP) per sqm	Total weight of this carpet 4,873g/m2. See Refs. 277 & 279.
Carpet tiles, nylon (Polyamide), pile weight 1100g/m2	378 MJ per sqm	19.7 (GWP) per sqm	19.7 (GWP) per sqm	Total weight of this carpet 5,123g/m2. See Refs. 277 & 279.
Polyethylterepthalate (PET)	106.50	5.56	—	Includes feedstock energy
Polypropylene	95.40	4.98	—	Includes feedstock energy, for per square meter see material profile
Polyurethane	72.10	3.76	—	Includes feedstock energy
Rubber	67.5 to 140	3.61 to 7.48	—	
Saturated Felt Underlay (impregnated with Asphalt or tar)	31.70	1.65	—	Ref. 94.

附录　建筑用基础材料碳排放数据库示例

续表

Materials	Embodied Energy & Carbon Coefficients			Comments
Wool	106.00	5.53	—	For per square meter see material profile. See Refs. 63, 201, 202 & 281 (Same author).
Cement				
General(UK weighted average)	4.5	0.73	0.74	Weighted average of all cement consumed within the UK. This includes all factory made cements (CEM Ⅰ, CEM Ⅱ, CEM Ⅲ, CEM Ⅳ) and further blending of fly ash and ground granulated blast furnace slag. This data has been estimated from the British Cement Association's factsheets (see Ref. 59). 23% cementitious additions on average.
Average CEM IPortland Cement, 94% Clinker	5.50	0.93	0.95	This is a standard cement with no cementitious additions (i.e. fly ash or blast furnace slag). Composition 94% clinker, 5% gypsum, 1% minor additional constituents (mac's). This data has been estimated from the British Cement Association's factsheets (see Ref. 59.).
6-20% Fly Ash (CEM Ⅱ/A-V)	5.28 to 4.51	0.88 (@ 6%) to 0.75 (@ 20%)	0.89 to 0.76	
21-35% Fly Ash (CEM Ⅱ/B-V)	4.45 to 3.68	0.74 to 0.61	0.75 to 0.62	See material profile for further details.
21-35% GGBS (CEM II/B-S)	4.77 to 4.21	0.76 to 0.64	0.77 to 0.65	
36-65% GGBS (CEM Ⅲ/A)	4.17 to 3.0	0.63 to 0.38	0.64 to 0.39	
66-80% GGBS (CEM Ⅱ/B)	2.96 to 2.4	0.37 to 0.25	0.38 to 0.26	
Fibre Cement Panels-Uncoated	10.40	1.09	—	Few data points. Selected data modified from Ref. 107.
Fibre Cement Panels-(Colour) Coated	15.30	1.28	—	
Mortar (1:3 cement:sand mix)	1.33	0.208	0.221	
Mortar (1:4)	1.11	0.171	0.182	Values estimated from the ICE Cement, Mortar & Concrete Model
Mortar (1:5)	0.97	0.146	0.156	
Mortar (1:6)	0.85	0.127	0.136	
Mortar (1:1/2:41/2 Cement:Lime:Sand mix)	1.34	0.200	0.213	

附录 建筑用基础材料碳排放数据库示例

续表

Materials	Embodied Energy & Carbon Coefficients			Comments
Mortar（1∶1∶6 Cement∶Lime∶Sand mix）	1.11	0.163	0.174	Values estimated from the ICE Cement, Mortar & Concrete Model
Mortar（1∶2∶9 Cement∶Lime∶Sand mix）	1.03	0.145	0.155	
Cement stabilised soil @ 5%	0.68	0.060	0.061	Assumed 5% cement content.
Cement stabilised soil @ 8%	0.83	0.082	0.084	Assumed 8% stabiliser contents (6% cement and 2% quicklime)
Ceramics				
General	10.00	0.66	0.70	Very large data range, difficult to select values for general ceramics.
Fittings	20.00	1.07	1.14	Ref. 1.
Sanitary Products	29.00	1.51	1.61	Limited data.
Tiles and Cladding Panels	12.00	0.74	0.78	Difficult to select, large range, limited data. See Ref. 292.
Clay				
General(Simple Baked Products)	3.00	0.23	0.24	General simple baked clay products (inc. terracotta and bricks)
Tile	6.50	0.45	0.48	
Vitrified clay pipe DN 100 & DN 150	6.20	0.44	0.46	
Vitrified clay pipe DN 200 & DN 300	7.00	0.48	0.50	
Vitrified clay pipe DN 500	7.90	0.52	0.55	
Concrete				
General	0.75	0.100	0.107	It is strongly recommended to avoid selecting a 'general' value for concrete. Selecting data for a specific concrete type (often a ready mix concrete) will give greater accuracy, please see material profile. Assumed cement content 12% by mass.
16/20 Mpa	0.70	0.093	0.100	
20/25 MPa	0.74	0.100	0.107	Using UK weighted average cement (more representative of 'typical' concrete mixtures).
25/30 MPa	0.78	0.106	0.113	
28/35 MPa	0.82	0.112	0.120	
32/40 MPa	0.88	0.123	0.132	
40/50 MPa	1.00	0.141	0.151	

附录 建筑用基础材料碳排放数据库示例

续表

Materials	Embodied Energy & Carbon Coefficients									Comments
% Cement Replacement-Fly Ash	0%	15%	30%	0%	15%	30%	0%	15%	30%	Note 0% is a concrete using a CEM I cement (not typical)
GEN 0 (6/8 MPa)	0.55	0.52	0.47	0.071	0.065	0.057	0.076	0.069	0.061	Compressive strength designation C6/8 Mpa. 28 day compressive strength under British cube method of 8 MPa, under European cylinder method 6 MPa. Possible uses: Kerb bedding and backing. Data is only cradle to factory gate but beyond this the average delivery distance of ready mix concrete is 8.3 km by road (see Ref. 244).
GEN 1 (8/10 MPa)	0.70	0.65	0.59	0.097	0.088	0.077	0.104	0.094	0.082	Possible uses: mass concrete, mass fill, mass foundations, trench foundations, blinding, strip footing.
GEN 2 (12/15 MPa)	0.76	0.71	0.64	0.106	0.098	0.087	0.114	0.105	0.093	—
GEN 3 (16/20 MPa)	0.81	0.75	0.68	0.115	0.105	0.093	0.123	0.112	0.100	Possible uses: garage floors.
RC 20/25 (20/25 MPa)	0.86	0.81	0.73	0.124	0.114	0.101	0.132	0.122	0.108	—
RC 25/30 (25/30 MPa)	0.91	0.85	0.77	0.131	0.121	0.107	0.140	0.130	0.115	Possible uses: reinforced foundations.
RC 28/35 (28/35 MPa)	0.95	0.90	0.82	0.139	0.129	0.116	0.148	0.138	0.124	Possible uses: reinforced foundations, ground floors.
RC 32/40 (32/40 MPa)	1.03	0.97	0.89	0.153	0.143	0.128	0.163	0.152	0.136	Possible uses: structural purposes, in situ floors, walls, superstructure.
RC 40/50 (40/50 MPa)	1.17	1.10	0.99	0.176	0.164	0.146	0.188	0.174	0.155	Possible uses: high strength applications, precasting.
PAV1	0.95	0.89	0.81	0.139	0.129	0.115	0.148	0.138	0.123	Possible uses: domestic parking and outdoor paving.
PAV2	1.03	0.97	0.89	0.153	0.143	0.128	0.163	0.152	0.137	Possible uses: heavy duty outdoor paving.
% Cement Replacement-Blast Furnace Slag	0%	25%	50%	0%	25%	50%	0%	15%	30%	Note 0% is a concrete using a CEM I cement
GEN 0 (6/8 MPa)	0.55	0.48	0.41	0.071	0.056	0.042	0.076	0.060	0.045	See fly ash mixtures
GEN 1 (8/10 MPa)	0.70	0.60	0.50	0.097	0.075	0.054	0.104	0.080	0.058	
GEN 2 (12/15 MPa)	0.76	0.62	0.55	0.106	0.082	0.061	0.114	0.088	0.065	
GEN 3 (16/20 MPa)	0.81	0.69	0.57	0.115	0.090	0.065	0.123	0.096	0.070	
RC 20/25 (20/25 MPa)	0.86	0.74	0.62	0.124	0.097	0.072	0.132	0.104	0.077	
RC 25/30 (25/30 MPa)	0.91	0.78	0.65	0.131	0.104	0.076	0.140	0.111	0.081	
RC 28/35 (28/35 MPa)	0.95	0.83	0.69	0.139	0.111	0.082	0.148	0.119	0.088	

附录 建筑用基础材料碳排放数据库示例

续表

Materials	Embodied Energy & Carbon Coefficients									Comments
RC 32/40 (32/40 MPa)	1.03	0.91	0.78	0.153	0.125	0.094	0.163	0.133	0.100	See fly ash mixtures
RC 40/50 (40/50 MPa)	1.17	1.03	0.87	0.176	0.144	0.108	0.188	0.153	0.115	
PAV1	0.95	0.82	0.70	0.139	0.111	0.083	0.148	0.118	0.088	
PAV2	1.03	0.91	0.77	0.153	0.125	0.094	0.163	0.133	0.100	

COMMENTS

The first column represents standard concrete, created with a CEM I Portland cement. The other columns are estimates based on a direct substitution of fly ash or blast furnace slag in place of the cement content. The ICE Cement, Mortar & Concrete Model was applied. Please see important notes in the concrete material profile.

REINFORCED CONCRETE-Modification Factors				
For reinforcement add this value to the appropriate concrete coefficient for each 100kg of rebar per m³ of concrete	1.04	0.072	0.077	Add for each 100kg steel rebar per m³ concrete. Use multiple of this value, i.e. for 150kg steel use a factor of 1.5 times these values.
EXAMPLE: Reinforced RC 25/30 MPa (with 110kg per m³ concrete)	1.92 MJ/kg (0.78+1.04*1.1)	0.185kg CO_2/kg (0.106+0.072*1.1)	0.198kg CO_2/kg (0.113+0.077*1.1)	with 110kg rebar per m³ concrete. UK weighted average cement. This assumes the UK typical steel scenario (59% recycled content). Please consider if this is in line with the rest of your study (goal and scope) or the requirements of a predefined method.

PRECAST (PREFABRICATED) CONCRETE-Modification Factors				
For precast add this value to the selected coefficient of the appropriate concrete mix	0.45	0.027	0.029	For each 1kg precast concrete. This example is using a RC 40/50 strength class and is not necessarily indicative of an average precast product. Includes UK recorded plant operations and estimated transportation of the constituents to the factory gate (38km aggregates, estimated 100km cement). Data is only cradle to factory gate but beyond this the average delivery distance of precast is 155km by road (see Ref. 244). UK weighted average cement. See also the new report on precast concrete pipes (Ref 300).
EXAMPLE: Precast RC 40/50 MPa	1.50 MJ/kg (1.00+0.50)	0.168kg CO_2/kg (0.141+0.027)	0.180kg CO_2/kg (0.151+0.029)	
EXAMPLE: Precast RC 40/50 with reinforcement (with 80kg per m³)	2.33 MJ/kg (1.50+1.04*0.8)	0.229kg CO_2/kg (0.171+0.072*0.8)	0.242kg CO_2/kg (0.180+0.077*0.8)	

CONCRETE BLOCKS (ICE CMC Model Values)				
Block-8 MPa Compressive Strength	0.59	0.059	0.063	Estimated from the concrete block mix proportions, plus an allowance for concrete block curing, plant operations and transport of materials to factory gate.
Block-10 MPa	0.67	0.073	0.078	
Block-12 MPa	0.72	0.082	0.088	
Block-13 MPa	0.83	0.100	0.107	

附录 建筑用基础材料碳排放数据库示例

续表

Materials	Embodied Energy & Carbon Coefficients			Comments
Autoclaved Aerated Blocks (AAC's)	3.50	0.24 to 0.375	—	Not ICE CMC model results.

NOMINAL PROPORTIONS METHOD (Volume), Proportions from BS 8500: 2006 (ICE Cement, Mortar & Concrete Model Calculations)

Materials	Embodied Energy & Carbon Coefficients			Comments
1 : 1 : 2 Cement : Sand : Aggregate	1.28	0.194	0.206	High strength concrete. All of these values were estimated assuming the UK average content of cementitious additions (i.e. fly ash, GGBS) for factory supplied cements in the UK, see Ref. 59, plus the proportions of other constituents.
1 : 1.5 : 3	0.99	0.145	0.155	Often used in floor slab, columns & load bearing structure.
1 : 2 : 4	0.82	0.116	0.124	Often used in construction of buildings under 3 storeys.
1 : 2.5 : 5	0.71	0.097	0.104	
1 : 3 : 6	0.63	0.084	0.090	Non-structural mass concrete.
1 : 4 : 8	0.54	0.069	0.074	

BY CEM I CEMENT CONTENT-kg CEM I cement content per cubic meter concrete (ICE CMC Model Results)

Materials	Embodied Energy & Carbon Coefficients			Comments
120kg/m³ concrete	0.49	0.060	0.064	
200kg/m³ concrete	0.67	0.091	0.097	Assumed density of 2,350kg/m³. Interpolation of the CEM I cement content is possible. These numbers assume the CEM I cement content (not the total cementitious content, i.e. they do not include cementitious additions). They may also be used for fly ash mixtures without modification, but they are likely to slightly underestimate mixtures that have additional GGBS due to the higher embodied energy and carbon of GGBS (in comparison to aggregates and fly ash).
300kg/m³ concrete	0.91	0.131	0.140	
400kg/m³ concrete	1.14	0.170	0.181	
500kg/m³ concrete	1.37	0.211	0.224	

附录 建筑用基础材料碳排放数据库示例

续表

Materials	Embodied Energy & Carbon Coefficients			Comments
MISCELLANEOUS VALUES				
Fibre-Reinforced	7.75 (?)	0.45 (?)	—	Literature estimate, likely to vary widely. High uncertainty.
Very High GGBS Mix	0.66	0.049	0.050	Data based on Lafarge 'Envirocrete', which is a C28/35 MPa, very high GGBS replacement value concrete
Copper				
EU Tube & Sheet	42.00	2.60	2.71	EU production data, estimated from Kupfer Institut LCI data. 37% recycled content (the 3 year world average). World average data is expected to be higher than these values.
Virgin	57.00	3.65	3.81	
Recycled	16.50	0.80	0.84	
Recycled from high grade scrap	18 (?)	1.1 (?)		Uncertain, difficult to estimate with the data available.
Recycled from low grade scrap	50 (?)	3.1 (?)		
Glass				
Primary Glass	15.00	0.86	0.91	Includes process CO_2 emissions from primary glass manufacture.
Secondary Glass	11.50	0.55	0.59	EE estimated from Ref 115.
Fibreglass (Glasswool)	28.00	1.54	—	Large data range, but the selected value is inside a small band of frequently quoted values.
Toughened	23.50	1.27	1.35	Only three data sources
Insulation				
General Insulation	45.00	1.86	—	Estimated from typical market shares. Feedstock Energy 16.5 MJ/kg (Included)
Cellular Glass	27.00	—	—	Ref. 54.
Cellulose	0.94 to 3.3	—	—	
Cork	4.00	0.19	—	Ref. 55.
Fibreglass (Glasswool)	28.00	1.35	—	Poor data difficult to select appropriate value
Flax (Insulation)	39.50	1.70	—	Ref. 2. 5.97 MJ/kg Feedstock Energy (Included)
Mineral wool	16.60	1.20	1.28	
Paper wool	20.17	0.63	—	Ref. 2

附录 建筑用基础材料碳排放数据库示例

续表

Materials	Embodied Energy & Carbon Coefficients			Comments
Polystyrene	See Plastics	See Plastics	—	see plastics
Polyurethane	See Plastics	See Plastics	—	see plastics
Rockwool	16.80	1.05	1.12	Cradle to Grave
Woodwool (loose)	10.80	—	—	Ref. 205.
Woodwool (Board)	20.00	0.98	—	Ref. 55.
Wool (Recycled)	20.90	—	—	Refs. 63, 201, 202 & 281.
Iron				
General	25.00	1.91 (?)	2.03	It was difficult to estimate the embodied energy and carbon of iron with the data available.
Lead				
General	25.21	1.57	1.67	Allocated (divided) on a mass basis, assumes recycling rate of 61%
Virgin	49.00	3.18	3.37	
Recycled	10.00	0.54	0.58	Scrap batteries are a main feedstock for recycled lead
Lime				
General	5.30	0.76	0.78	Embodied carbon was difficult to estimate
Linoleum				
General	25.00	1.21	—	Data difficult to select, large data range.
Miscellaneous				
Asbestos	7.40	—	—	Ref. 4.
Calcium Silicate Sheet	2.00	0.13	—	Ref. 55.
Chromium	83	5.39	—	Ref. 22.
Cotton, Padding	27.10	1.28	—	Ref. 38.
Cotton, Fabric	143	6.78	—	Ref. 38.
Damp Proof Course/Membrane	134 (?)	4.2 (?)	—	Uncertain estimate.
Felt General	36	—	—	
Flax	33.50	1.70	—	Ref. 2.
Fly Ash	0.10	0.008	—	No allocation from fly ash producing system.
Grit	0.12	0.01	—	Ref. 114.
Ground Limestone	0.62	0.032	—	
Carpet Grout	30.80	—	—	Ref. 169.
Glass Reinforced Plastic-GRP-Fibreglass	100	8.10	—	Ref. 1.
Lithium	853	5.30	—	Ref. 22.

附录　建筑用基础材料碳排放数据库示例

续表

Materials	Embodied Energy & Carbon Coefficients			Comments
Mandolite	63	1.40	—	Ref. 1.
Mineral Fibre Tile (Roofing)	37	2.70	—	Ref. 1.
Manganese	52	3.50	—	Ref. 22.
Mercury	87	4.94	—	Ref. 22.
Molybedenum	378	30.30	—	Ref. 22.
Nickel	164	12.40	—	Ref. 114.
Perlite-Expanded	10.00	0.52	—	Ref. 114.
Perlite-Natural	0.66	0.03	—	Ref. 114.
Quartz powder	0.85	0.02	—	Ref. 114.
Shingle	11.30	0.30	—	Ref. 70.
Silicon	2355	—	—	Ref. 167.
Slag (GGBS)	1.60	0.083	—	Ground Granulated Blast Furnace Slag (GGBS), economic allocation.
Silver	128.20	6.31	—	Ref. 148.
Straw	0.24	0.01	—	Refs. 63, 201, 202 & 281.
Terrazzo Tiles	1.40	0.12	—	Ref. 1.
Vanadium	3710	228	—	Ref. 22.
Vermiculite-Expanded	7.20	0.52	—	Ref. 114.
Vermiculite-Natural	0.72	0.03	—	Ref. 114.
Vicuclad	70.00	—	—	Ref. 1.
Water	0.01	0.001	—	
Wax	52.00	—	—	Ref. 169.
Wood stain/Varnish	50.00	5.35	—	Ref. 1.
Yttrium	1470	84.00	—	Ref. 22.
Zirconium	1610	97.20	—	Ref. 22.
Paint				
General	70.00	2.42	2.91	Large variations in data, especially for embodied carbon. Includes feedstock energy. Water based paints have a 70% market share. Water based paint has a lower embodied energy than solvent based paint.
EXAMPLE: Single Coat	10.5 MJ/Sqm	0.36kg CO_2/Sqm	0.44	Assuming 6.66 Sqm Coverage per kg
EXAMPLE: Double Coat	21.0 MJ/Sqm	0.73kg CO_2/Sqm	0.87	Assuming 3.33 Sqm Coverage per kg
EXAMPLE: Triple Coat	31.5 MJ/Sqm	1.09kg CO_2/Sqm	1.31	Assuming 2.22 Sqm Coverage per kg

附录 建筑用基础材料碳排放数据库示例

续表

Materials	Embodied Energy & Carbon Coefficients			Comments
Waterborne Paint	59.00	2.12	2.54	Waterborne paint has a 70% of market share. Includes feedstock energy.
Solventborne Paint	97.00	3.13	3.76	Solventborne paint has a 30% share of the market. Includes feedstock energy. It was difficult to estimate carbon emissions for Solventborne paint.
Paper				
Paperboard (General for construction use)	24.80	1.29	—	Excluding calorific value (CV) of wood, excludes carbon sequestration/biogenic carbon storage.
Fine Paper	28.20	1.49	—	Excluding CV of wood, excludes carbon sequestration
EXAMPLE: 1 packet A4 paper	70.50	3.73	—	Standard 80g/sqm printing paper, 500 sheets a pack. Doesn't include printing.
Wallpaper	36.40	1.93	—	
Plaster				
General (Gypsum)	1.80	0.12	0.13	Problems selecting good value, inconsistent figures, West et al believe this is because of past aggregation of EE with cement
Plasterboard	6.75	0.38	0.39	See Ref [WRAP] for further info on GWP data, including disposal impacts which are significant for Plasterboard.
Plastics	Main data source: PlasticsEurope (www.plasticseurope.org) ecoprofiles			
General	80.50	2.73	3.31	35.6 MJ/kg Feedstock Energy (Included). Determined by the average use of each type of plastic used in the European construction industry.
ABS	95.30	3.05	3.76	48.6 MJ/kg Feedstock Energy (Included)
General Polyethylene	83.10	2.04	2.54	54.4 MJ/kg Feedstock Energy (Included). Based on average consumption of types of polyethylene in European construction
High Density Polyethylene (HDPE) Resin	76.70	1.57	1.93	54.3 MJ/kg Feedstock Energy (Included). Doesn't include the final fabrication.

附录 建筑用基础材料碳排放数据库示例

续表

Materials	Embodied Energy & Carbon Coefficients			Comments
HDPE Pipe	84.40	2.02	2.52	55.1 MJ/kg Feedstock Energy (Included)
Low Density Polyethylene (LDPE) Resin	78.10	1.69	2.08	51.6 MJ/kg Feedstock Energy (Included). Doesn't include the final fabrication
LDPE Film	89.30	2.13	2.60	55.2 MJ/kg Feedstock Energy (Included)
Nylon (Polyamide) 6 Polymer	120.50	5.47	9.14	38.6 MJ/kg Feedstock Energy (Included). Doesn't include final fabrication. Plastics Europe state that two thirds of nylon is used as fibres (textiles, carpets …etc) inEurope and that most of the remainder as injection mouldings. Dinitrogen monoxide and methane emissions are very significant contributors to GWP.
Nylon (polyamide) 6,6 Polymer	138.60	6.54	7.92	50.7 MJ/kg Feedstock Energy (Included). Doesn't include final fabrication (i.e. injection moulding). See comments for Nylon 6 polymer.
Polycarbonate	112.90	6.03	7.62	36.7 MJ/kg Feedstock Energy (Included). Doesn't include final fabrication.
Polypropylene, Orientated Film	99.20	2.97	3.43	55.7 MJ/kg Feedstock Energy (Included).
Polypropylene, Injection Moulding	115.10	3.93	4.49	54 MJ/kg Feedstock Energy (Included). If biomass benefits are included the CO_2 may reduce to 3.85kg CO_2/kg, and GWP down to 4.41kg CO_2e/kg.
Expanded Polystyrene	88.60	2.55	3.29	46.2 MJ/kg Feedstock Energy (Included)
General Purpose Polystyrene	86.40	2.71	3.43	46.3 MJ/kg Feedstock Energy (Included)
High Impact Polystyrene	87.40	2.76	3.42	46.4 MJ/kg Feedstock Energy (Included)
Thermoformed Expanded Polystyrene	109.20	3.45	4.39	49.7 MJ/kg Feedstock Energy (Included)
Polyurethane Flexible Foam	102.10	4.06	4.84	33.47 MJ/kg Feedstock Energy (Included). Poor data availability for feedstock energy
Polyurethane Rigid Foam	101.50	3.48	4.26	37.07 MJ/kg Feedstock Energy (Included). Poor data availability for feedstock energy

附录 建筑用基础材料碳排放数据库示例

续表

Materials	Embodied Energy & Carbon Coefficients			Comments
PVC General	77.20	2.61	3.10	28.1 MJ/kg Feedstock Energy (Included). Based on market average consumption of types of PVC in the European construction industry
PVC Pipe	67.50	2.56	3.23	24.4 MJ/kg Feedstock Energy (Included). If biomass benefits are included the CO_2 may reduce to 2.51kg CO_2/kg, and GWP down to 3.23kg CO_2e/kg.
Calendered Sheet PVC	68.60	2.61	3.19	24.4 MJ/kg Feedstock Energy (Included). If biomass benefits are included the CO_2 may reduce to 2.56kg CO_2/kg, and GWP down to 3.15kg CO_2e/kg.
PVC Injection Moulding	95.10	2.69	3.30	35.1 MJ/kg Feedstock Energy (Included). If biomass benefits are included the CO_2 may reduce to 2.23kg CO_2/kg, and GWP down to 2.84kg CO_2e/kg.
UPVC Film	69.40	2.57	3.16	25.3 MJ/kg Feedstock Energy (Included)
Rubber				
General	91.00	2.66	2.85	40 MJ/kg Feedstock Energy (Included)
Sand				
General	0.081	0.0048	0.0051	Estimated from real UK industrial fuel consumption data
Sealants and adhesives				
Epoxide Resin	137.00	5.70	—	42.6 MJ/kg Feedstock Energy (Included). Source: www.plasticseurope.org
Mastic Sealant	62 to 200	—	—	
Melamine Resin	97.00	4.19	—	Feedstock energy 18 MJ/kg-estimated from Ref 34.
Phenol Formaldehyde	88.00	2.98	—	Feedstock energy 32 MJ/kg-estimated from Ref 34.
Urea Formaldehyde	70.00	2.76	—	Feedstock energy 18 MJ/kg-estimated from Ref 34.
Soil				
General (Rammed Soil)	0.45	0.023	0.024	
Cement stabilised soil @ 5%	0.68	0.060	0.061	Assumed 5% cement content.

附录 建筑用基础材料碳排放数据库示例

续表

Materials	Embodied Energy & Carbon Coefficients			Comments
Cement stabilised soil @ 8%	0.83	0.082	0.084	Assumed 8% stabiliser content (6% cement and 2% lime).
GGBS stabilised soil	0.65	0.045	0.047	Assumed 8% stabiliser content (8% GGBS and 2% lime).
Fly ash stabilised soil	0.56	0.039	0.041	Assumed 10% stabiliser content (8% fly ash and 2% lime).
Steel	Main data source: International Iron & Steel Institute (IISI) LCA studies (www.worldsteel.org)			
UK (EU) STEEL DATA-EU average recycled content-See material profile (and Annex on recycling methods) for usage guide				
General-UK (EU) Average Recycled Content	20.10	1.37	1.46	EU 3-average recycled content of 59%. Estimated from UK's consumption mixture of types of steel (excluding stainless). All data doesn't include the final cutting of the steel products to the specified dimensions or further fabrication activities. Estimated from World Steel Association (Worldsteel) LCA data.
Virgin	35.40	2.71	2.89	
Recycled	9.40	0.44	0.47	Could not collect strong statistics on consumption mix of recycled steel.
Bar & rod-UK (EU) Average Recycled Content	17.40	1.31	1.40	EU 3-average recycled content of 59%
Virgin	29.20	2.59	2.77	
Recycled	8.80	0.42	0.45	
Coil (Sheet)-UK (EU) Average Recycled Content	18.80	1.30	1.38	Effective recycled content because recycling route is not typical. EU 3-average recycled content of 59%
Virgin	32.80	2.58	2.74	
Recycled	Not TypicalProduction Route			
Coil (Sheet), Galvanised-UK (EU) Average Recycled Content	22.60	1.45	1.54	Effective recycled content because recycling route is not typical. EU 3-average recycled content of 59%
Virgin	40.00	2.84	3.01	
Engineering steel-Recycled	13.10	0.68	0.72	
Pipe-UK (EU) Average Recycled Content	19.80	1.37	1.45	Effective recycled content because recycling route is not typical. EU 3-average recycled content of 59%

续表

Materials	Embodied Energy & Carbon Coefficients			Comments
Virgin	34.70	2.71	2.87	
Recycled	Not Typical Production Route			
Plate-UK (EU) Average Recycled Content	25.10	1.55	1.66	Effective recycled content because recycling route is not typical. EU 3-average recycled content of 59%
Virgin	45.40	3.05	3.27	
Recycled	Not Typical Production Route			
Section-UK (EU) Average Recycled Content	21.50	1.42	1.53	
Virgin	38.00	2.82	3.03	
Recycled	10.00	0.44	0.47	
Wire-Virgin	36.00 (?)	2.83 (?)	3.02	
Stainless	56.70	6.15		World average data from the Institute of Stainless Steel Forum (ISSF) life cycle inventory data. Selected data is for the most popular grade (304). Stainless steel does not have separate primary and recycled material production routes.
OTHER STEEL DATA-'R.O.W' and 'World' average recycled contents-See material profile (and Annex on recycling methods) for usage guide				
General-R.O.W. Avg. Recy. Cont.	26.20	1.90	2.03	Rest of World (non-E.U.) consumption of steel. 3 year average recycled content of 35.5%.
General-World Avg. Recy. Cont.	25.30	1.82	1.95	Whole world 3 year average recycled content of 39%.
Bar & rod-R.O.W. Avg. Recy. Cont.	22.30	1.82	1.95	Comments above apply. See material profile for further information.
Bar & rod-World Avg. Recy. Cont.	21.60	1.74	1.86	
Coil-R.O.W. Avg. Recy. Cont.	24.40	1.81	1.92	
Coil-World Avg. Recy. Cont.	23.50	1.74	1.85	
Coil, Galvanised-R.O.W. Avg. Recy. Cont.	29.50	2.00	2.12	
Coil, Galvanised-World Avg. Recy. Cont.	28.50	1.92	2.03	
Pipe-R.O.W. Avg. Recy. Cont.	25.80	1.90	2.01	

附录 建筑用基础材料碳排放数据库示例

续表

Materials	Embodied Energy & Carbon Coefficients			Comments
Pipe-World Avg. Recy. Cont.	24.90	1.83	1.94	Comments above apply. See material profile for further information.
Plate-R.O.W. Avg. Recy. Cont.	33.20	2.15	2.31	
Plate-World Avg. Recy. Cont.	32.00	2.06	2.21	
Section-R.O.W. Avg. Recy. Cont.	28.10	1.97	2.12	
Section-World Avg. Recy. Cont.	27.10	1.89	2.03	
Stone	Data on stone was difficult to select, with high standard deviations and data ranges.			
General	1.26 (?)	0.073 (?)	0.079	ICE database average (statistic), uncertain. See material profile.
Granite	11.00	0.64	0.70	Estimated from Ref 116.
Limestone	1.50	0.087	0.09	Estimated from Ref 188.
Marble	2.00	0.116	0.13	
Marble tile	3.33	0.192	0.21	Ref. 40.
Sandstone	1.00 (?)	0.058 (?)	0.06	Uncertain estimate based on Ref. 262.
Shale	0.03	0.002	0.002	
Slate	0.1 to 1.0	0.006 to 0.058	0.007 to 0.063	Large data range
Timber	Note: These values were difficult to estimate because timber has a high data variability. These values exclude the energy content of the wooden product (the Calorific Value (CV) from burning). See the material profile for guidance on the new data structure for embodied carbon (i.e. split into foss and bio)			
General	10.00	$0.30_{fos}+0.41_{bio}$	$0.31_{fos}+0.41_{bio}$	Estimated from UK consumption mixture of timber products in 2007 (Timber Trade Federation statistics). Includes 4.3 MJ bio-energy. All values do not include the CV of timber product and exclude carbon storage.
Glue Laminated timber	12.00	$0.39_{fos}+0.45_{bio}$	$0.42_{fos}+0.45_{bio}$	Includes 4.9 MJ bio-energy.
Hardboard	16.00	$0.54_{fos}+0.51_{bio}$	$0.58_{fos}+0.51_{bio}$	Hardboard is a type of fibreboard with a density above 800kg/m^3. Includes 5.6 MJ bio-energy.
Laminated Veneer Lumber	9.50	$0.31_{fos}+0.32_{bio}$	$0.33_{fos}+0.32_{bio}$	Ref 150. Includes 3.5 MJ bio-energy.
MDF	11 (?)	$0.37_{fos}+0.35_{bio}$	$0.39_{fos}+0.35_{bio}$	Wide density range (350-800kg/m^3). Includes 3.8 MJ bio-energy.

附录 建筑用基础材料碳排放数据库示例

续表

Materials	Embodied Energy & Carbon Coefficients			Comments
Oriented Strand Board (OSB)	15.00	$0.42_{fos}+0.54_{bio}$	$0.45_{fos}+0.54_{bio}$	Estimated from Refs.103 and 150. Includes 5.9 MJ bio-energy.
Particle Board	14.50	$0.52_{fos}+0.32_{bio}$	$0.54_{fos}+0.32_{bio}$	Very large data range, difficult to select appropriate values. Modified from CORRIM reports. Includes 3.2 MJ bio-energy (uncertain estimate).
Plywood	15.00	$0.42_{fos}+0.65_{bio}$	$0.45_{fos}+0.65_{bio}$	Includes 7.1 MJ bio-energy.
Sawn Hardwood	10.40	$0.23_{fos}+0.63_{bio}$	$0.24_{fos}+0.63_{bio}$	It was difficult to select values for hardwood, the data was estimated from the CORRIM studies (Ref.88). Includes 6.3 MJ bio-energy.
Sawn Softwood	7.40	$0.19_{fos}+0.39_{bio}$	$0.20_{fos}+0.39_{bio}$	Includes 4.2 MJ bio-energy.
Veneer Particleboard (Furniture)	$23_{(fos+bio)}$	(?)	(?)	Unknown split of fossil based and biogenic fuels.
Tin				
Tin Coated Plate (Steel)	19.2 to 54.7	1.04 to 2.95	—	
Tin	250.00	13.50	14.47	lack of modern data, large data range
Titanium				
Virgin	361 to 745	19.2 to 39.6 (??)	20.6 to 42.5 (??)	lack of modern data, large data range, small sample size
Recycled	258.00	13.7 (??)	14.7 (??)	lack of modern data, large data range, small sample size
Vinyl Flooring				
General	68.60	2.61	3.19	23.58 MJ/kg Feedstock Energy (Included), Same value as PVC calendered sheet. Note: *the book version of ICE contains the wrong values. These values are up to date*
Vinyl Composite Tiles (VCT)	13.70	—	—	Ref.94.
Zinc				
General	53.10	2.88	3.09	Uncertain carbon estimates, currently estimated from typical UK industrial fuel mix. Recycled content of general Zinc 30%.
Virgin	72.00	3.90	4.18	
Recycled	9.00	0.49	0.52	
Miscellaneous (No material profiles):				
	Embodied Energy-MJ	Embodied Carbon-Kg CO_2		

附录 建筑用基础材料碳排放数据库示例

续表

Materials	Embodied Energy & Carbon Coefficients			Comments
PV Modules	MJ/sqm	Kg CO_2/sqm		
Monocrystalline	4750（2590 to 8640）	242（132 to 440）	—	Embodied carbon estimated from typical UK industrial fuel mix. This is not an ideal method.
Polycrystalline	4070（1945 to 5660）	208（99 to 289）	—	
Thin Film	1305（775 to 1805）	67（40 to 92）	—	
Roads	Main data source：ICE reference number 147			
Asphalt road-Hot construction method-40 yrs	2,509 MJ/Sqm	93kg CO_2/Sqm	99kg CO_2/Sqm	730 MJ/Sqm Feedstock Energy (Included). For more detailed data see reference 147. (Swedish study). The data in this report was modified to fit within the ICE framework. Includes all sub-base layers to construct a road. Sum of construction, maintenance, operation.
Construction	1,069 MJ/Sqm	30.9kg CO_2/Sqm	32.8kg CO_2/Sqm	480 MJ/Sqm Feedstock Energy (Included)
Maintenance-40 yrs	471 MJ/Sqm	11.6kg CO_2/Sqm	12.3kg CO_2/Sqm	250 MJ/Sqm Feedstock Energy (Included)
Operation-40 yrs	969 MJ/Sqm	50.8kg CO_2/Sqm	54.0kg CO_2/Sqm	Swedish scenario of typical road operation, includes street and traffic lights (95% of total energy), road clearing, sweeping, gritting and snow clearing.
Asphalt road-Cold construction method-40 yrs	3,030 MJ/Sqm	91kg CO_2/Sqm	97kg CO_2/Sqm	1,290 MJ/kg Feedstock Energy (Included). Sum of construction, maintenance, operation.
Construction	825 MJ/Sqm	26.5kg CO_2/Sqm	28.2kg CO_2/Sqm	320 MJ/Sqm Feedstock Energy (Included)
Maintenance-40 yrs	1,556 MJ/Sqm	13.9kg CO_2/Sqm	14.8kg CO_2/Sqm	970 MJ/Sqm Feedstock Energy (Included)
Operation-40 yrs	969 MJ/Sqm	50.8kg CO_2/Sqm	54.0kg CO_2/Sqm	See hot rolled asphalt.
Concrete road-40 yrs	2,084 MJ/Sqm	142kg CO_2/Sqm	—	Sum of construction, maintenance, operation.
Construction	885 MJ/Sqm	77kg CO_2/Sqm	—	
Maintenance-40 yrs	230 MJ/Sqm	14.7kg CO_2/Sqm	—	
Operation-40 yrs	969 MJ/Sqm	50.8kg CO_2/Sqm	—	Swedish scenario of typical road operation, includes street and traffic lights (95% of total energy), and also road clearing, sweeping, gritting and snow clearing.

附录 建筑用基础材料碳排放数据库示例

续表

Materials	Embodied Energy & Carbon Coefficients			Comments

Note: *The above data for roads were based on a single reference (ref 145). There were other references available but it was not possible to process the reports into useful units (per sqm). One of the other references indicates a larger difference between concrete and asphalt roads than the data above. If there is a particular interest in roads the reader is recommended to review the literature in further detail.*

Windows	MJ per Window			
1.2m×1.2m Single Glazed Timber Framed Unit	286 (?)	14.6 (?)	—	Embodied carbon estimated from typical UK industrial fuel mix
1.2m×1.2m Double Glazed (Air or Argon Filled):	—	—	—	
Aluminium Framed	5470	279	—	
PVC Framed	2150 to 2470	110 to 126	—	
Aluminium-Clad Timber Framed	950 to 1460	48 to 75	—	
Timber Framed	230 to 490	12 to 25	—	
Krypton Filled Add:	510	26	—	
Xenon Filled Add:	4500	229	—	

NOTE: *Not all of the data could be converted to full GHG's. It was estimated from the fuel use only (i.e. Not including any process related emissions) the full CO_2 e is approximately 6 percent higher than the CO_2 only value of embodied carbon. This is for the average mixture of fuels used in the UK industry.*

Annex A: Boundary Conditions

The boundaries within the ICE database are cradle-to-gate. However even within these boundaries there are many possible variations that affect the absolute boundaries of study. One of the main problems of utilising secondary data resources is variable boundaries since this issue can be responsible for large differences in results. The ICE database has its ideal boundaries, which it aspires to conform to in a consistent manner. However, with the problems of secondary data resources there may be some instances where modification to these boundaries was not possible. The ideal boundaries are listed below:

Item	Boundaries treatment
Delivered energy	All delivered energy is converted into primary energy equivalent, see below.
Primary energy	Default method, traced back to the 'cradle'.
Primary electricity	Included, counted as energy content of the electricity (rather than the opportunity cost of energy).
Renewable energy (inc. electricity)	Included.
Calorific Value (CV)/Heating value of fossil fuel energy	Default values are Higher Heating Values (HHV) or Gross Calorific Values (GCV), both are equivalent metrics.
Calorific value of organic fuels	Included when used as a fuel, excluded when used as a feedstock, e.g. timber offcuts burnt as a fuel include the calorific value of the wood, but timber used in a table excludes the calorific value of the wooden product.

附录 建筑用基础材料碳排放数据库示例

续表

Item	Boundaries treatment
Feedstock energy	Fossil fuel derived feedstocks are included in the assessment, but identified separately. For example, petrochemicals used as feedstocks in the manufacture of plastics are included. See above category for organic feedstock treatment.
Carbon sequestration and biogenic carbon storage	Excluded, but ICE users may wish to modify the data themselves to include these effects.
Fuel related carbon dioxide emissions	All fuel related carbon dioxide emissions which are attributable to the product are included.
Process carbon dioxide emissions	Included; for example CO_2 emissions from the calcination of limestone in cement clinker manufacture are counted.
Other greenhouse gas emissions	The newest version of the ICE database (2.0) has been expanded to include data for GHGs. The main summary table shows the data in CO_2 only and for the GHGs in $CO_2 e$.
Transport	Included within specified boundaries, i.e. typically cradle-to-gate.

混凝土 EPD 评价采用的 PCR 示例

TABLE OF CONTENT

Guidance document ··· 113
General Introduction ··· 116
0   General information ··· 117
1   Scope ·· 117
   1.1   General ·· 117
   1.2   Objective of this PCR ·· 118
   1.3   Types of EPD with respect to life cycle stages covered ······························· 119
   1.4   Comparability of EPD for construction products ·· 121
   1.5   Additional information provided in the EPD ·· 123
   1.6   Ownership, responsibility and liability for the EPD ····································· 124
   1.7   Communication formats ·· 124
2   Product category rules for LCA ··· 125
   2.1   The product category covered by this PCR ·· 125
   2.2   Life cycle stages included ··· 125
   2.3   Calculation rules for the LCA ··· 130
   2.4   Inventory analysis ··· 146
   2.5   Types of EPD ·· 151
   2.6   Other Aspects ·· 152
   2.7   Impact assessment ··· 154
3   Content of the EPD ·· 158
   3.1   Declaration of general information ··· 158
   3.2   Declaration of environmental parameters derived from LCA ······················ 159
   3.3   Scenarios and additional technical information ·· 166
   3.4   Aggregation of information modules ··· 171
4   Project Verification report ·· 172
   4.1   General ·· 172
   4.2   LCA-related elements of the Project Verification Report ····························· 172
   4.3   Documentation on additional information ·· 174
   4.4   Data availability for verification ··· 174
   4.5   Verification and validity of an EPD ·· 175
5   Terms and definitions ··· 176
6   Abbreviations ··· 188

GUIDANCE DOCUMENT

INTRODUCTION

This Guidance document is intended to be read in conjunction with the Product Category Rules for Unreinforced Concrete produced for the WBCSD Cement Sustainability Initiative. Its aim is to explain the Product Category Rules (PCR) to a general audience and provide hypothetical examples of how it should be used in the concrete industry to aid its implementation.

HOW IS THIS DOCUMENT LAID OUT?

The Guidance is interspersed with the PCR, and highlighted in GREEN.
The PCR is based on EN 15804: 2012. Where text has been taken directly from the version of the standard circulated for Final Enquiry, this text has been included in italics.
Terms and definitions are provided in Section 5, and Abbreviations are found in Section 6.

WHAT ARE ENVIRONMENTAL PRODUCT DECLARATIONS (EPD)?

An Environmental Product Declaration (EPD) is a Type III environmental declaration as defined by ISO 14025: 2006. It is a voluntary document that provides data in a predefined format, including information about the environmental impacts associated with the manufacturing of a product or system.
EPD are issued by a Program Operator, following rules known as Product Category Rules (PCR), which ensure that products are assessed in a consistent manner.

HOW IS AN EPD PRODUCED?

In order to produce an EPD, a Life Cycle Assessment (LCA) must be undertaken. An LCA is an evaluation of the environmental impacts of a product, which considers all the relevant stages in its life cycle, from extraction of raw materials (the "cradle") to the disposal at the end of the product's useful life (the "grave"). For intermediate products and final products where many scenarios may apply, the later life cycle stages are often excluded from the assessment and the resulting scope is referred to as "cradle to gate" or "cradle to site".
The first step in completing an LCA is to obtain manufacturer's data on their inputs (materials and energy) and outputs (products, wastes and emissions to air, water and land) of a process. This data is used to produce an LCA model, which when linked with LCA data for the "upstream" and "downstream" processes such as raw materials manufacture,

energy production or waste treatment, produce a Life Cycle Inventory, which is a full list of the resource use and environmental emissions or environmental burdens.

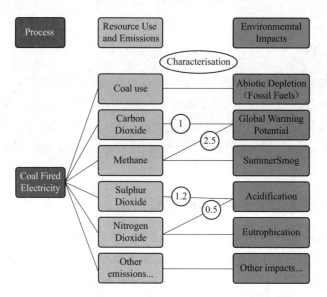

These environmental burdens are then assigned to impact categories by applying technical or scientific coefficients known as characterisation factors. An example of this process is shown in Figure 0 1 for some impact categories. For instance, methane has 25 times the potential of CO_2 for Global Warming. The total impact of a product's production can be calculated by summing the impacts of all emissions and resource uses. The number of impact categories provided by an EPD can vary depending on the PCR but this PCR assesses seven core environmental impact categories.

HOW RIGOROUS ARE EPD?

In order to meet the ISO 14025: 2006 requirements, an EPD must give independently verified, quantitative information on the environmental effects of a product, based on data provided by the manufacturer. All EPD follow the ISO 14040 framework, an international standard that defines the underlying concepts of Life Cycle Assessment. EPD are verified by an external independent examiner, who has knowledge of LCA and the specific products. EPD aim to capture at least 98% of impacts, and should be representative of a year's production. It is accepted that there will be variation for a product produced at an individual time or factory compared to that annual average. Differences for key indicator results between EPD for products of less than 10% where there is limited shared background information cannot generally be considered to provide a definitive comparison, due to the uncertainties associated with the different upstream data and assumptions made for the products. However, for similar products, where background data is shared (for example the same upstream data for electricity, cement and aggregates is used), then much smaller differences between products can be considered relevant.

HOW CAN EPD BE USED?

The Environmental Product Declaration is a standardised document, which can used to communicate the potential environmental impacts associated with a product, but EPD may only be used to compare the environmental properties of several products or systems if they have the same function in the building or other construction works and are based on the same PCR-this is a requirement of the ISO standard for EPD. Particular attention should be paid to the declared unit of the EPD, as well as the properties of the product within the building or other construction works (insulation, strength, etc), in order to make a relevant comparison.

An EPD is essential for sustainable building certifications, for instance according to:

- DGNB-German Sustainable Building Certification Scheme: a building level LCA must be carried out, and IBU EPD can be used and are included in various calculation tools. Credits are given for building materials with low environmental impacts.
- BREEAM-UK Environmental Building Certification Scheme: uplift credits are available for products qualified by independently verified EPD.
- LEED-US Environmental Building Certification Scheme: Pilot Credits are available for products with LCA-based assessments-the most credits are available for independently verified, manufacturer-specific EPD.

GENERAL INTRODUCTION

The International EPD® System is based on a hierarchic approach following the international standards:
- ISO 9001, Quality management systems
- ISO 14001, Environmental management systems
- ISO 14040, LCA-Principles and procedures
- ISO 14044, LCA-Requirements and guidelines
- ISO 14025, Type III environmental declarations
- ISO 21930, Environmental declaration of building products
- EN 15804, Sustainability of construction works—Environmental product declarations

The General programme Instructions are based on these standards, as well as instructions for developing Product Category Rules (PCR).

The documentation to The International EPD® System includes three separate parts (www.environdec.com):
- Introduction, intended uses and key programme elements
- General Programme Instructions Supporting annexes
- Supporting annexes

This PCR document specifies further and adds additional minimum requirements on EPDs of the product group defined below complementary to the above mentioned general requirement documents. Principle programme elements concerning the Product Category Rules (PCR) included in The International EPD® System are presented below.

PURPOSE	ELEMENT IDENTIFICATION AND PRINCIPAL APPROACH
Complying with principles set in ISO 14025 on modularity and comparability	1. "Book-keeping LCA approach" 2. A Polluter-Pays (PP), allocation method
Simplifying work to develop Product Category Rules (PCR)	3. PCR Module Initiative (PMI) in order to structure PCR in modules according to international classification 4. PCR moderator for leadership and support of the PCR work
Secure international participation in PCR work	5. Global PCR Forum for open and transparent EPD stakeholder consultation
Facilitating, identification and collection of LCA-based information	6. Selective data quality approach for specific and generic data

Product Category Rules (PCR) are specified for specified information modules "gate-to-gate", so called core modules. The structure and aggregation level of the core modules are defined by the United Nation Statistics Division-Classification Registry CPC codes (http://unstats.un.org). The PCR also provides rules for which methodology and data to use in the full LCA, i.e. life cycle parts up-streams and down-streams the core module.

The PCR also has requirements on the information given in the EPD, e.g. additional environmental information. A general requirement on the information in the EPD is that all information given in the EPD, mandatory and voluntary, shall be verifiable.

In the EPD, the environmental performance associated with each of the three life-cycle stages mentioned above are reported separately.

0　GENERAL INFORMATION

Date:	2013-02-12
Registration no:	PCR 2013: 02
This PCR was prepared by:	WBCSD Cement Sustainability Initiative
Appointed PCR moderator:	Jane Anderson, PE International, j.anderson@pe-international.com
Open consultation period:	2012-09-11 until 2012-10-23
Valid within the following geographical representativeness:	Global
Valid until:	2018-02-12
More information on this PCR's website:	http://environdec.com/en/Product-Category-Rules/Detail/?Pcr=8108

1　SCOPE

1.1　GENERAL

Product Category Rules (PCR) are the underlying rules which are used to produce an Environmental Product Declaration. Environmental Product Declarations are known as EPD, and are an example of a Type III Environmental Label as defined in ISO Standard ISO 14020: 2001, Environmental Labels and Declarations-General principles.

An EPD is a document which provides consistent, robust and independently verified data about the environmental impacts and aspects of a product or service. EPD are based on a process known as Life Cycle Assessment or LCA, and are governed by a number of International and European standards including ISO 14025: 2006 for Type III EPD, ISO 21930: 2007 for EPD for construction products and EN 15804, a European standard providing core rules for Product Category Rules for Construction Products which will be published in 2012. This PCR is intended to be compliant with all of these standards mentioned and has been based on EN 15804.

The PCR are the rules which specify in detail how a particular product category, such as unreinforced concrete, should be considered, to ensure that EPD for this product category are considered and calculated in the same way, and the results can therefore be considered as consistent.

This PCR covers the Product Category "unreinforced concrete", and can be applied to any unreinforced concrete product, including ready mix concrete, and precast concrete products such as blocks.

The PCR:

- defines the parameters, such as environment impact indicators (such as global warming potential) and inventory indicators (such as net freshwater consumption) to be declared and the way in which they are collated (i. e. -how they are considered and calculated) and reported,
- describes the stages of a concrete product's life cycle (such as extraction or manufacture or installation) which are considered in the EPD and which processes are to be included within each stage,
- defines rules for the provision of additional information about the product to enable further life cycle stages (such as transport to site or disposal) to be assessed,
- includes the rules (such as the system boundary, cut-offs and allocation rules) for calculating the Life Cycle Inventory (LCI)-this is the list of resources, outputs and emissions to air, water and land associated with the product and the Life Cycle Impact Assessment (LCIA)-the environmental impacts resulting from the inventory) underlying the EPD, including the specification of the data quality (the age, representativeness and geography of datasets for upstream and downstream processes) to be applied,
- defines the conditions under which construction products can be compared based on the information provided by EPD.

EPD are intended to enable building or construction works level evaluation, and EPD can only be used to make product to product comparisons when the effect of the products at the building or construction works level, over the life cycle, has been considered. As a result, product to product comparisons using cradle to gate or cradle to site EPD can only be made in very limited circumstances, which are set out in the PCR.

The rules for reporting environmental and health information in relation to emissions to air, water and land from the product in use, the effect on "indoor air quality" and "run-off" or "eluate" are not yet defined, measured and reported in an agreed manner and standards are still being developed. As these issues relate to the phase of the life cycle after the product has been installed, at present, this is an area where the PCR is expected to develop in the future.

1.2 OBJECTIVE OF THIS PCR

This PCR is a set of specific rules, requirements and guidelines for developing Type Ⅲ environmental declarations, known as Environmental Product Declarations (EPD) for the product category of unreinforced concrete.

An EPD according to this PCR will provide consistent and quantified environmental information for an unreinforced concrete product on a harmonized and scientific basis. This PCR is intended to be compliant with EN 15804 and therefore cradle to gate EPD produced in accordance with this PCR should be usable within Europe within any building or construction works level environmental assessment compliant with EN 15804 and/or EN 15978.

Europe has developed a suite of environmental standards through Technical Committee TC350. These include EN 15804 which sets out how to develop Product Category Rules for construction products and EN 15978 which sets out how to undertake building level assessment. These standards have been developed to overcome the barriers to trade which have been experienced with the use of various incompatible national EPD schemes. At the time of writing, the use of EPD and building or construction works level assessment is most advanced in Europe, and for this reason, EN 15804 has been chosen as the underlying standard. Work is on-going in North America and Australasia for example developing PCRs and building level assessments, and at present, we understand that these are looking to use EN 15804 as an underlying standard.

The purpose of an EPD in the construction sector is to provide the basis for assessing buildings and other construction works, and to assist in identifying those construction products which cause less stress to the environment considering the whole building life cycle.

Thus the objective of this PCR is to ensure:

- *the provision of verifiable and consistent data for an EPD, based on LCA;*
- *that comparisons between construction products are carried out in the context of their application in the building;*
- *the communication of the environmental information of construction products from business to business.*

Declarations based on this PCR are not comparative assertions.

A comparative assertion is an environmental claim regarding the superiority or equivalence of one product versus a competing product that performs the same function. An EPD does not make any statements that the product covered by the EPD is better or worse than any other product.

An EPD produced according to this PCR can be used to assess the impact of a building or construction works, using the data from many EPD in order to "build" the impact of the building or construction works.

1.3 TYPES OF EPD WITH RESPECT TO LIFE CYCLE STAGES COVERED

The LCA based information in an EPD provided by this PCR covers:

- The product stage. A product EPD covers raw material supply, transport, manufacturing and associated processes; this EPD is said to be "cradle to gate".

Using terminology from EN 15804 a cradle to gate EPD uses information modules A1 to A3. An information module is EPD data for a specific life cycle phase and product, which can be used with other information modules for the product to cover the full life cycle, and can be combined with information modules for other products to provide information for product systems or building or construction works.

Additionally, EPD provided by this PCR may include LCA based information modules covering two further life cycle stages, known as "cradle to gate with options":
- the transport to construction site stage (Information module A4 from EN 15804)
- the construction stage (Information module A5 from EN 15804).

The Cradle to gate stage must be included in any EPD, and should remain the same for a given manufacturing location, irrespective of where the product is used. The transport to construction site and construction stage modules may vary depending on where the product is delivered and how it is used in the building or construction works. For this reason, these information modules A4 and A5 are based on a scenario, for example, (e.g. the typical situation, or a common situation), and are optional, as they may not be appropriate for many customers as their scenario may be based on using the same product in a different situation.

EPD for other products or using other PCRs may provide LCA based information covering additional life cycle stages as follows:
- The product stage and selected further life cycle stage including the use phase (information modules B1 to B7 from EN15804), and end-of-life (information modules C1 to C4 from EN 15804). Although outside the system boundary, EPD may also provide information on the benefits of recycling and reuse beyond the system boundary (Information module D from EN 15804);
- The full life cycle of a product according to the system boundary (see 2.3.4). In this case the EPD covers the product manufacturing stage, transportation to the construction site, installation into the building or construction works, use and maintenance, replacements, demolition, waste processing for re-use, recovery, recycling and disposal, and disposal. Such a system boundary is said to be 'cradle to grave' and becomes an EPD of construction products based on a LCA, (i.e. covering all information modules A1 to C4). In a cradle to grave EPD, the benefits of recycling beyond the system boundary (information module D) may also be included.

An information module may contain: the values of the pre-determined parameters and the technical information underlying their quantification, relevant technical information for further calculation of the environmental performance, scenarios for further calculation of the environmental performance.

Guidance Examples: It is possible to have an EPD for a substance or preparation (e. g. cement), for a product (e. g. concrete block, concrete kerb or concrete slab), for a construction service (e. g. cleaning service as part of maintenance) and, for an assembly of products, and/or a construction element (e. g. wall).

1.4 COMPARABILITY OF EPD FOR CONSTRUCTION PRODUCTS

In principle the comparison of products on the basis of their EPD is defined by the contribution they make to the environmental performance of the building. Consequently, comparison of the environmental performance of construction products using the EPD information shall be based on the product's use in and its impacts on the building, and shall consider the complete life cycle of the product within the building or construction works.

The Cradle to Gate EPD as covered by this PCR are not provided in the context of a building or construction works and cannot be used to compare construction products and construction services unless the considerations below are met.

Comparisons are possible at the sub-building level, e. g. for assembled systems, components, or for products covering the cradle to gate life cycle stage if it is ensured that the comparison considers the products in the context of the entire building or construction works-this shall be maintained by ensuring that:

- *the same functional requirements as defined by legislation or in the client's brief are met, and*
- *the environmental performance and technical performance of any assembled systems, components, or products excluded are the same, and*
- *the amounts of any material excluded are the same, and*
- *the processes and impacts of the excluded life cycle stages are the same, and*
- *the influence of the product systems on the operational aspects and impacts of the building are taken into account.*

The information provided for such comparison shall be transparent to allow the purchaser or user to understand the limitations of comparability.

The client for the building or construction works will have provided a brief-this may say that they require a specific functionality for a particular part of the building or construction works-for example, that a meeting room needs a given acoustic performance and

> that the building needs to meet Passivhaus standards. In this case, the internal walls of the room will need to meet a given acoustic performance, but not a specific thermal performance. The external walls will need to meet both an acoustic and a thermal performance. Two concrete blocks, with different thermal performance, but which both met the client's brief in terms of acoustic performance, could be compared for use in the internal walls.
> The difference between two products may appear significant at the product level, but may be insignificant in the building or construction works context.

As the EPD covered by this PCR do not cover all life cycle stages, then investigations will be required to determine the environmental aspects and impacts of specific scenarios for the impacts beyond the cradle to gate life cycle stage. These calculations shall be based on scenarios and conditions that are appropriate for the building or construction works as the object of assessment.

> Guidance: Examples are provided below where two concrete blocks can or cannot be compared directly taking into account the requirements above:
> Two concrete blocks, with different thermal resistances and/or thermal mass, used in an external wall: in this case, the two blocks may affect the amount of heat required for the building over its life in different ways and this would need to be taken into consideration for any comparison. Alternatively, additional materials to match the performance of the two products (if a particular performance is required from the client's brief) would need to be included in the comparison.
> Two concrete blocks with different thermal resistance used in an internal wall: in this case, the difference in functionality would not alter the thermal performance of the building and the products could therefore be compared.
> Two concrete floor slabs with different densities: in this case, the heavier floor slab may require stronger or deeper foundations, and this would need to be taken into consideration in any comparison of the two products.
> Two concrete blocks, one using thin joint mortar, the other conventional: in this instance, the block and mortar would need to be considered as a system and the two systems would need to be compared.
> Two concretes, one using expanded polystyrene (EPS) as an aggregate: these two concretes would have different End of Life Scenarios owing to their different constituents and this would need to be included in any comparison.

1.5 ADDITIONAL INFORMATION PROVIDED IN THE EPD

This PCR allows the provision of additional technical information which has not been derived from LCA and that forms part of the EPD by providing a basis for the development of scenarios. Such Additional Technical information describes technical conditions that could be used to develop scenarios and characterise the product's technical and functional performance during the excluded life cycle stages "Transport to site", "Construction", "Use" and the "End of life", for any scenario based calculations of the LCA based parameters. See also 4.2.

Two examples of the type of additional information that could be provided to enable the subsequent development of LCA for scenarios are included below.

Additional Information	Typical transport to site for X Ltd's concrete blocks in UK
Fuel type and consumption of vehicle or vehicle type used for transport e.g. long distance truck, boat etc.	Road, 28 tonne GLW, Diesel Litres/100km, EURO Class
Distance	120km
Capacity utilisation (including empty returns)	70%
Bulk density of transported products	1800kg/m^3
Volume capacity utilisation factor (factor: = 1or<1or≥1for compressed or nested packaged products)	1

Example of additional information on Transport to Construction Site Scenario:

Additional Information	Typical Installation per m^3 concrete
Ancillary materials for installation (specified by material);	Formwork: 1m^2 18mm plywood, 95% reuse rate
Net consumption of fresh water for installation	0.7m^3
Other resource use for installation	n/a
Quantitative description of energy type (regional mix) and consumption during the installation process	Compaction: 0.2MJ/m^3
Wastage of materials on the construction site before waste processing, generated by the product's installation (specified by type)	Product: 0.3kg Formwork (plywood) 0.05m^2 18mm plywood
Output materials (specified by type) as result of waste processing at the construction site e.g. of collection for recycling, for energy recovery, disposal (specified by route)	Product-reuse on site Formwork-energy recycling
Direct emissions to ambient air, soil and water	Any water used evaporates or is discharged to sewer.

The rules for reporting environmental and health information in relation to emissions to air, water and land from the product in use, the effect on "indoor air quality" and "run-off" or "eluate" are not yet defined, measured and reported in an agreed manner and standards are still being developed. At present therefore, the PCR does not provide for the provision of additional information regarding the "in use" life cycle stage, but this is an area where the PCR is expected to develop in the future.

> European Standards Technical Committee CEN/TC 351 will be developing European standards to cover the testing and assessment of emissions from products to air, water and land during use, and once in place, these will be used within a future update of EN 15804 to report on these emissions during the use phase. These would be reported in Module B1 (from EN 15804), and additional information could be included in these modules to enable results to be provided for specific scenarios.
> An additional aspect of information which could be covered in this part of the EPD is carbonation of concrete products. When poured concrete is curing, it reabsorbs some CO_2 from the atmosphere. Re-absorption is however small compared to the emissions from cement production. More CO_2 is absorbed throughout the lifetime of the concrete product, but very slowly. A recent study (ECRA 2008) has summarized the available published research results with respect to concrete recarbonation. Due to the present lack of accurate and quantifiable data, recarbonisation of cement and concrete is currently not part of this PCR. This may change in the future.

1.6　OWNERSHIP, RESPONSIBILITY AND LIABILITY FOR THE EPD

A manufacturer or a group of manufacturers producing an EPD to this PCR are the sole owners and have liability and responsibility for their EPD.

> Guidance: Environmental information provided in the EPD is based on manufacturer declared data. The verifier will check the general plausibility of this data but the responsibility for providing data which is an accurate representation of the company's operations lies with the company. Manufacturers should be aware that competitors may challenge environmental data if they believe that the EPD is not based upon accurate data. All data provided as the basis of the EPD should be capable of substantiation and be verifiable if possible.

1.7　COMMUNICATION FORMATS

Any communication format of the EPD produced according to this PCR shall be in accordance with EN 15942, Sustainability of construction works—Environmental product declarations—Communication formats: business to business.

2 PRODUCT CATEGORY RULES FOR LCA

2.1 THE PRODUCT CATEGORY COVERED BY THIS PCR

The product category referred to in this PCR covers unreinforced concrete products for use in buildings and other construction works, including the following products:

- ready mixed concrete
- concrete blocks, but excluding aircrete
- concrete kerbstones
- mortar.

It does not cover reinforced concrete including fibre cement. For the purposes of this PCR, concrete is defined as "Material formed by mixing cement, coarse and fine aggregate and water, with or without the incorporation of admixtures or addition, which develops its properties by the hardening of the cement paste (cement and water)".

In accordance with the standard EN 206-1: 2001, concrete is classified by:

- Compressive strength class.

For any compressive strength class, concrete must be defined by:

- Environmental exposure class.
- Slump class (optional).

> Guidance Examples:
> A concrete without reinforcement or embedded metal: in all exposures except where there is freeze/thaw, abrasion or chemical attack: may have a range of compressive strengths, for example of C12/15 or C30/37 and environmental exposure class X0.
> A concrete used where it is exposed to significant attack from freeze-thaw cycles whilst wet, for example where it would be exposed to high water saturation, without de-icing agents, such as a horizontal concrete surface exposed to rain and freezing, may, for example, have a compressive strength of C30/37 and an environmental exposure class XF3.

2.2 LIFE CYCLE STAGES INCLUDED

2.2.1 GENERAL

The environmental information of EPD covered by this PCR covers the life cycle stage ("cradle to gate"), and optionally, the transport to site and construction stages.

The cradle to gate life cycle stage must be provided in the EPD and this, can, if required, be broken down into 3 life cycle stages using terminology from EN 15804:
- A1, raw material extraction and processing, processing of secondary material input (e. g. recycling processes)
- A2, transport to the manufacturer
- A3, manufacturing, including impacts from direct energy generation and waste disposal related to the manufacturing process

Module A1, A2 and A3 may be declared as one aggregated module A1-3.

> Guidance: A1 is associated with "upstream" processes using common LCA terminology, whereas A3 covers "gate to gate".

Information modules include impacts and aspects related to losses in the module in which the losses occur (i. e. production, transport, and waste processing and disposal of the waste from a process are included in the module in which the waste is produced).

> Guidance: For a concrete manufacturer, the splitting of the cradle to gate stage into the three sub-stages may be useful to demonstrate that the major impact comes from the raw materials rather than the processes undertaken by the concrete manufacturer. Note that electricity generation impacts related to electricity use by a concrete manufacturer would be included within stage A1 as they are impacts which occur upstream of manufacture, whereas the impacts of burning fuels such as natural gas used by the manufacturer would be included within the manufacturing stage A3.
>
> Concrete producers should note that they do have the ability to alter the impacts of the product stage A1 through decisions relating to mix design or/and the amount and type of electricity used for example.

2.2.2 PRODUCT STAGE A1

The product stage A1 includes, with the exception of ancillary materials and packaging used in the studied product manufacturing process, and transport of inputs to the studied manufacturing process, all extraction, energy production and manufacturing processes which occur upstream of the studied product manufacturing process, including:
- Extraction and processing of raw materials (e. g. mining processes) and biomass production and processing (e. g. agricultural or forestry operations) used as input for manufacturing the product;
- Extraction and processing of primary fuels used as input for manufacturing the product;
- Reuse of products or materials from a previous product system used as input for manufacturing the product, but not including those processes that are part of the waste pro-

cessing of the previous product system until it reaches the end-of-waste state (see 2.3.4.5);
- Processing of secondary materials used as input for manufacturing the product, but not including those processes that are part of the waste processing in the previous product system until it reaches the end-of-waste state;
- Generation of electricity, steam and heat used in the product manufacturing process, which have been generated offsite, also including their extraction, refining and transport;
- Energy recovery and other recovery processes from secondary fuels, that are used as input for manufacturing the product, but not including those processes that are part of waste processing in the previous product system until it reaches the end-of-waste state;
- Processing up to the end-of-waste state (see 2.3.4.5) and disposal of any final residues produced during any process stage included in A1;
- any transport of raw materials within the upstream supply chain, apart from the delivery of materials to the studied manufacturing process.

> Guidance: Impacts from the extraction and manufacturing of ancillary materials and packaging materials are excluded from product stage A1 and are included in Product Stage A3-this is based on the text of EN 15804.
> Impacts from transport of all materials and fuel to the studied manufacturing process are included in product stage A2.
> Example: For a concrete manufacturer, the impacts of the cement production process, including the extraction of limestone, transport of limestone to the cement manufacturer and the extraction, processing and use of energy in cement production would be included in A1. The impacts of electricity used by the concrete manufacturer, including extraction of fuels, electricity production and distribution, would be included in A1. The impacts of natural gas used by the concrete manufacturer, including extraction of gas, processing and distribution, and emissions from the combustion of gas would be included in A3. The transport of the cement to the concrete manufacturer would be included in A2.

2.2.3 PRODUCT STAGE A2

The product stage A2 includes all transport processes upstream and during the manufacturing process, but excluding transport of waste from the manufacturing process. Extraction and fuel processing impacts associated with transport would be included in this stage.
- A2 Transportation up to the factory gate and internal transport;

混凝土 EPD 评价采用的 PCR 示例

> Guidance: Transport of manufacturing waste is included in Product stage A3.

Impacts of transport should be based on the typical delivery of products, taking account of the mode of transport (road, rail, water, air), vehicle type, fuel and efficiency, load, distance travelled and empty return journey. This will be most significant for the bulk materials used (aggregates) and the materials which come a considerable distance.

Product stage A3

Product stage A3 includes all manufacturing processes for the studied product, plus manufacturing of ancillary materials and pre-products, and packaging, plus any waste processing and transport of wastes to the end-of-waste state or disposal.

- A3 Production of ancillary materials or pre-products;
- A3 Manufacturing of products and co-products, including the combustion of any primary fuels used in the manufacturing process;
- A3 Manufacturing of Packaging;
- A1-A3 processing up to the end-of-waste state (see 2.3.4.5) or disposal of final residues including any packaging not leaving the factory gate with the product.

> Guidance: The output of waste during this life cycle stage may reach the end-of-waste state when it complies with the conditions described in 2.3.4.5, end-of-waste state. They are then allocated as co-products as 2.3.5.3.
> Product stage A3 most closely aligns with the Scope 1 emissions recorded by the Greenhouse Gas Protocol, however it covers all impacts, not just global warming/climate change. As with Scope 1, Product stage A3 does not include the upstream impacts from extracting and processing primary fuels, such as coal and natural gas-these are included in A1. The impacts associated with any electricity, heat, steam or secondary fuels used in manufacturing are included in Product Stage A1.
> It should be noted that the scope of A1, A2 and A3 vary through the supply chain-for a limestone quarry, A3 would cover the extraction of the stone; for a cement plant, limestone extraction would be included in A1, transport to the plant in A2 and the cement kiln operation in A3; for a ready mix concrete plant, cement production would be included in A1, and A3 would just cover the mixing operation.
> In order to simplify reporting of the results, each EPD can report the impact for each category as pre factory and factory impacts. This can be made by reporting the product stage A1 and A2 together (as pre-factory impact), and the product stage A3 separately (as factory impacts).

2.2.4 CONSTRUCTION STAGE-TRANSPORT TO SITE A4

The construction-transport to site process stage can be reported within the EPD based on the typical situation, and includes the optional information modules for:
- Transportation from the production gate to the construction site;
- Storage of products during the distribution phase from product gate to construction site, including the provision of heating, cooling, humidity control, etc. ;
- Wastage of construction products (covering the relevant processes required to manufacture the material lost through the wastage of products);
- Waste Processing of the waste from product packaging and product wastage which occurs during the transport and distribution processes up to the end-of-waste state (see 2.3.4.5) or disposal of final residues;
- Washing of vehicles if this is undertaken at a different location from the production site or the construction site.

> Guidance: Impacts of transport should be based on the typical delivery of products, taking account of the mode of transport (road, rail, water, air), vehicle type, fuel and efficiency, load, distance travelled and empty return journey. This will be most significant for the bulk materials used (aggregates) and the materials which come a considerable distance.
>
> Where, for example, concrete is wasted during the delivery operation-for example where it is over-ordered or significant concrete residues need to be cleaned from the lorry, then the manufacture of the material which is wasted, and the disposal of the material which is wasted need to be included within product stage A4.
>
> Cleaning of vehicles which occurs at the concrete production factory would be included in Product stage A3. Cleaning of vehicles which occurs at the construction site would be included in Product Stage A5.

2.2.5 CONSTRUCTION STAGE-ON SITE PROCESSES A5

The Construction Stage-on site processes may be reported within the EPD based on the typical situation. If this stage is included in the EPD, it may include the optional information modules for:
- Storage of products, including the provision of heating, cooling, humidity control, etc. ;
- Washing of vehicles undertaken on the construction site;
- The pumping of concrete to place it on site;
- The cooling or heating of the concrete or the site to enable the concrete to cure;
- The use of any water provided on site to mix the concrete or to enable the concrete to cure;

- The net consumption of any formwork taking into account normal rates of reuse or recycling;
- Any ancillary materials required for releasing concrete from formwork;
- Any cleaning of equipment or formwork required;
- Wastage of construction products on site (covering the relevant processes required to manufacture the material lost through the wastage of products);
- Waste Processing of the waste from product packaging and product wastage during the construction processes up to the end-of-waste state (see 2.3.4.5) or disposal of final residues;
- Installation of the product into the building or construction works including manufacture and transportation of ancillary materials and any energy or water required for installation or operation of the construction site. It also includes on-site operations to the product.

Guidance: Data for general operations of the construction site (eg. site huts, scaffolding etc) should be considered at the building or construction works level and do not need to be included in the provision of a product level EPD. However, processes which directly relate to the concrete, for example, processes required to compact the concrete, provide a smooth surface (e.g. floating) or other processes required for the curing of the concrete should be modelled within this construction stage.

2.3 CALCULATION RULES FOR THE LCA

2.3.1 FUNCTIONAL UNIT

A functional unit, which needs to reflect the required functional performance of the product within the building or construction works over the full life cycle, can only be used for a cradle to grave EPD. The cradle to gate and cradle to site EPD covered by this PCR are based on a declared unit, although this includes some aspects of functionality.

Guidance: At the point where concrete leaves the ready-mix plant or arrives at the construction site for example, it is still possible for the concrete to be used in a wide variety of different functions in the building or construction works-for example foundations, floor slabs, road pavement or pathways. In each of these functions, the service life and exposure of the concrete might be different, and potentially the end of life scenario for the concrete may also vary. Similarly, a concrete block could be used in an internal or external wall, with thin joint or normal mortar-both of these options will impact on the thermal performance of the block in the building or construction works. For Cradle to Gate and Cradle to Site EPD, the function of the product within the building or construction works is too uncertain and a functional unit, which needs to include information on the required functional performance of the product within the building or construction works over the full life cycle, is therefore inappropriate.

2.3.2 DECLARED UNIT

The declared unit is used instead of the functional unit when the precise function of the product or scenarios for the product at the building level is not stated or is unknown. The declared unit provides a reference by means of which the material flows of the information module of a construction product are normalised (in a mathematical sense) to produce data, expressed on a common basis. It provides the reference for combining material flows attributed to the construction product. The declared unit shall relate to the typical applications of products.

Within this PCR, it is relevant for the products below to use the following declared units incorporating relevant aspects of functionality related to the potential purpose in the building or construction works to enable more useful and consistent comparison at the building or construction works level:

- Concrete: $1m^3$ of concrete with a given compressive strength class, environmental exposure class as per EN 206 or an equivalent stated national standard as relevant to the potential use of the product in the building or construction works (density shall be specified);
- Concrete Blocks: 1 block with a given strength (dimensions and density to be specified);
- Concrete Blockwork: $1m^2$ of blockwork with a given strength (dimensions and density shall be specified);
- Pipework: 1 metre of unreinforced pipework with a given capacity, (dimensions and mass shall be specified) (pipes are normally reinforced at diameters greater than 600 mm);
- Mortar: 1 cubic metre of concrete with a given strength (density shall be specified).

Guidance: see, for example, http://www.icjonline.com/views/POV_VRK.pdf for a review of national standards covering environmental exposure classes.

Thermal resistance, thermal mass, acoustic performance and other relevant performance information may also be relevant to the description of the declared unit.

For the development of e.g. transport and disposal scenarios, conversion factors to mass per declared unit shall always be provided.

If any non-mass based unit is used, then information such as density must be provided to allow the impact for 1 kg of product to be calculated.

Guidance: Reasons for declaring units other than those listed include the need to use units normally used for design, planning, procurement and sale.

2.3.3 RELATING DATA TO UNIT PROCESS AND DECLARED UNITS

An appropriate flow shall be determined for each unit process. The quantitative input and output data of the unit process shall be calculated in relation to this flow. Based on the flow chart and the flows between unit processes, the flows of all unit processes are related to the reference flow. The calculation should result in all system input and output data being referenced to the declared unit for the product.

> Guidance: The reference flow is a given amount of the product covered by the EPD. This will be set by choosing the declared unit for the product and may be, for example, $1m^3$ of ready mix concrete. For this product, we can consider the various processes (unit processes) which are used to provide inputs and treat outputs of the product system. An input unit process may be the production of cement-the flow for this unit process would be measured in kg, and the mass of cement used in the product system would be used to multiply the impacts associated with the production of 1 kg of cement by the amount of cement used. This is repeated for all the flows throughout the whole product system so that all impacts are accounted for in relation to the reference flow, the declared unit for the product.

2.3.4 SYSTEM BOUNDARIES

2.3.4.1. General

LCA is conducted by defining product systems as models describing the key elements of physical systems. The system boundary defines the unit processes to be included in the system model.

This clause specifies the boundary of the product system under study and in particular the boundary with any previous or subsequent product systems in the life of a building. This is most particularly relevant in setting the boundary between processes producing waste and processes using the waste.

Any approach must ensure that the system boundaries are transparent, well defined and applicable to any construction product. They must also be consistent, to ensure that if the system producing waste and the system using waste were both modelled, that there would be no double counting or undercounting of impacts.

The setting of the system boundaries follows these two principles:

- The "modularity principle": Where processes influence the product's environmental performance during its life cycle, they shall be assigned to the module of the life cycle where they occur; all environmental aspects and impacts are declared in the life cycle stage where they appear;
- The "polluter pays principle": Processes of waste processing shall be assigned to the product system that generates the waste until the end-of-waste state is reached (see 2.3.4.5).

Guidance:

The "modularity" principle: For example, construction product packaging is disposed of on the construction site-the impacts of disposal must therefore be included in module A5 (construction site).

The "polluter pays principle": The system boundary must be set so that it is clear which impacts belong in which systems. For example, when a process, Process 1, creates waste, which is then recycled and used in another process, Process 2, the system boundary can be set at any point between the production of the waste and the use of the recycled material. If it is set at the production of waste, then all the impacts of recycling go to Process 2 using the recycled material, and none to Process 1, the waste producer. If it is set at the point where the recycled material is used, then all the impacts go to Process 1, and none to Process 2. Other system boundaries could be the point at which the output moves from a waste to a recycled product or the point at which Process 1 stops paying for waste treatment and Process 2 starts to pay for it. Another option is to take the impact of the waste treatment and recycling process, and to split the impact 50: 50 (or some other split) between the two processes.

The system boundary needs to be defined between processes and nature, between processes and their wastes (pre-consumer recycling), and between processes and any post-consumer recycling at the end of life.

2.3.4.2. Boundary of the technosphere with nature

The system boundary with nature is set to include those processes that provide the material and energy inputs into the system and the following manufacturing, and transport processes up to the factory gate as well as the emissions to air, soil and water and the processing of any waste arising from those processes.

The time period over which inputs to and outputs from the system shall be accounted for is 100 years from the year for which the data set is deemed representative.

Guidance: Where materials are extracted from nature (mining, quarrying, water extraction etc) the boundary of life cycle assessment is normally taken at the first point at which humans have an influence on the environment. For this reason, processes such as land clearance, removal of overburden etc need to be considered within the scope of the LCA, although their impact will be spread over all the material that is extracted as a result. When emissions are returned to nature, for example, emissions to air are released to the atmosphere, emissions to water are released to a river or lake and materials are released to land, then their impacts arising from this release to nature are tracked for a 100 year period.

> For waste placed in landfill, then this system is not considered "nature", but the emissions from the landfill system over 100 years, for example from off-gassing, landfill gas recovery processes, leachate treatment, and any other emissions associated with the landfill will be considered.
> Some LCA will take account of the emissions over 60,000 years from deposition in the landfill. This assumes that the landfill linings and systems will have collapsed, and that all material will return to nature. This timescale is not considered within this PCR.

2.3.4.3. Boundary between product systems

Where secondary materials, or energy recovered from secondary fuels are used, the system boundary between the system under study and the previous system (providing the secondary materials) is set where outputs of the previous system, e.g. wastes, by-products, end-of-life material or waste energy, reach the end-of-waste state (see 2.3.4.5).

For waste flows leaving the system under study and entering another systemX (e.g. because it is recycled or energy is recovered), the system boundary is also set at the end-of-waste state. This ensures consistency in methodology.

> Guidance: some by-products or co-products may reach the end-of-waste state as soon as they are produced. In this situation, the system boundary is immediately after their production, and some allocation of impacts from the producing system to the receiving system may be required.
> Example: A manufacturer may produce metal waste, for example which reaches the end-of-waste status as soon as it has been collected together. In this situation, it can be treated as a co-product with value, and impact can be allocated to this metal for recycling based on its value at the end-of-waste, compared to the value of other products produced.

2.3.4.4. System boundary for energy recovery processes

In some circumstances, energy recovery from waste may occur within the process under study. In this instance, the end-of-waste state is theoretically reached at some point within the energy recovery process, but within this PCR, the system boundary is set at the point that the wastes enter the energy recovery process. Any impacts which can be physically assigned to the energy recovery process, for example specific emissions from the combustion of the waste, can be separately identified within the EPD, but will be included within the impacts provided.

Guidance Example: Where a waste, such as whole tyres, is used for energy recovery within a cement kiln, the impacts of all waste processes (such as collection and transport) are within the automotive system until the whole tyres are put into the cement kiln. Non-biogenic CO_2 emissions from the tyres (which can be calculated from the non-biogenic carbon content of the tyres) can be recorded as a subset of the total CO_2 emissions within the LCI, and hence as a subset of the Global Warming Potential impact within the EPD.

2.3.4.5. End-of-waste state

Waste produced by a system reaches the end-of-waste state when it complies with all of the following criteria as shown in Figure 3-1 overleaf:

- the recovered material, product or construction element is commonly used for specific purposes;
- a market or demand, identified e.g. by a positive economic value, exists for such a recovered material, product or construction element;
- the recovered material, product or construction element fulfils the technical requirements for the specific purposes and meets the existing legislation and standards applicable to products; and
- the use of the recovered material, product or construction element will not lead to overall adverse environmental or human health impacts.

Guidance: The "specific purpose" in this context is not restricted to the function of a certain product but can also be applied to a material serving as input to the production process of another product or of energy.

The criterion for "overall adverse environmental or human health impacts" shall refer to the limit values for pollutants set by regulations in place at the time of assessment and where necessary and shall take into account adverse environmental effects. The presence of any hazardous substances exceeding these limits in the waste or showing one or more properties as listed in existing applicable legislation, e.g. in the European Waste Framework Directive 2008/98/EC, prevents the waste from reaching the end-of-waste state.

Guidance: http://publications.environment-agency.gov.uk/PDF/GEHO0411BTRD-E-E.pdf is guidance which has been published in the UK by the relevant National agency to explain how a waste can be determined as hazardous within the European Framework Directive, and is useful in providing sources to definitions of what are dangerous substances and the relevant threshold values and testing mechanisms for these. Please not any references to national legislation within the document are only relevant in the UK however.

混凝土 EPD 评价采用的 PCR 示例

This definition of the end-of-waste state is to be used in Europe and for products sold in Europe, and in any other location where there are no local rules. In other locations where the national rules differ, the local approach to end-of-waste state should be applicable.

> Guidance Example: say a waste is produced in a country outside Europe which has different waste legislation, the criteria provided in 2.3.4.5 do not need to be used to define the end-of-waste state but the national rules can be used to define the end-of-waste state of the waste.
>
> For use of waste or secondary material as a fuel, impacts from the burning of the fuel are included in the system whether the fuel is a considered a waste or to have reached the end-of-waste state, although emissions from waste can be separately identified.
>
> In this situation, the total reported impacts will be the same irrespective of the national rules. To ensure consistency with EN15804 for the treatment of secondary fuels which may be defined as a waste in different jurisdictions, in the regions where the fuels are not defined as waste, non-biogenic CO_2 emissions will be included in the total indicator. In the regions where they are considered waste for energy recovery i.e alternative fuels, non-biogenic CO_2 emissions will be included in the system boundary and can be reported as a sub-total of the total indicator. However, the energy for both secondary fuels and waste for energy recovery will be accounted for as "use of non-renewable secondary fuels".
>
> To ensure consistency with EN15804 for the treatment of input materials which may be defined as a waste or secondary material in different jurisdictions, in the regions where the materials are not defined as waste, allocation of impacts from the producing system to these inputs will be based on economic value. In the regions where the inputs are considered waste, allocation will not be considered and they will not have impacts from the producing system

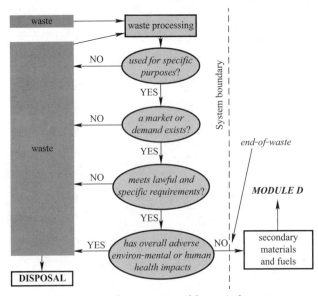

Figure 2 1: Decision-tree of for end-of-waste

> Guidance: Having a positive economic value is only one criterion by which the end-of-waste state is judged. It is therefore possible for a substance to have a positive economic value but still be a waste, for example if it is considered hazardous.

2.3.5 ALLOCATION OF INPUT FLOWS AND OUTPUT EMISSIONS

2.3.5.1. General

Many industrial processes produce not just the intended product but co-products and by-products. Normally the material flows of inputs are not distributed between them in a simple way. Intermediate and discarded products can be recycled to become inputs for other processes. When dealing with systems involving multiple products and recycling processes, allocation should be avoided whenever possible. Where unavoidable, allocation should be considered carefully and the allocation method chosen should be justified.

In this PCR, the rules for allocation are based on the guidance given in EN 15804 6.5.3, which is based on the guidance in EN ISO 14044; however, the basic procedures and assumptions used in EN ISO 14044 have been refined in order to reflect the goal and scope of EN 15804 and EN 15643-2.

The principle of modularity shall be maintained. Where processes influence the product's environmental performance during its life cycle, they shall be assigned to the life cycle stage where they occur.

The sum of the allocated inputs and outputs of a unit process shall be equal to the inputs and outputs of the unit process before allocation. This means no double counting or omission of inputs or outputs through allocation is permitted.

> Guidance: Products and functions are the outputs and/or services provided by the process, having a positive economic value.
>
> In industrial processes there are often a wide variety of different types of materials produced in conjunction with the intended product. In business vocabulary, these are normally identified as by-products, co-products, intermediate products, non-core products or sub-products. In this PCR these terms are treated as being equivalent. However for the allocation of environmental aspects and impacts a distinction between co-products and products is made in this PCR.

2.3.5.2. Co-product allocation

Co-products are any outputs of any product manufacturing system which leave the system boundary. These can be co-products which are intended to be made, co-products which are not intended to be made or any wastes from manufacturing which following processing reach the end-of-waste state are considered to leave the product system (see 2.3.5.3). The impact of processing any waste output from the system must be included in the system until it reaches the end-of-waste state and leaves the system boundary. If an output never reaches the end-of-waste state, then its waste processing and disposal remains within the system boundary of the product. If it is not possible to consider the production of the main product on its own, in order to decide the impact associated with the product, and co-product (s) at the system boundary, the impact of the system needs to be allocated between them.

> Guidance: In the context of coal fired electricity production, the main output is electricity, and other outputs from the system may include fuel ash, furnace bottom ash, flue gas desulphurization gypsum or recovered heat for example. In this context, electricity is the product, and the other outputs are co-products if they reach the end-of-waste state, or are never considered wastes.

Allocation shall be avoided as far as possible by dividing the unit process to be allocated into different sub-processes that can be allocated to the co-products and by collecting the input and output data related to these sub-processes.

- If a process can be sub-divided but respective data are not available, the inputs and outputs of the system under study should be partitioned between its different products or functions in a way which reflects the underlying physical relationships between them; i.e. they shall reflect the way in which the inputs and outputs are changed by quantitative changes in the products or functions delivered by the system;

In the case of joint co-production, where the processes cannot be sub-divided, allocation shall respect the main purpose of the processes studied, allocating all relevant products and functions appropriately. The purpose of a plant and therefore of the related processes is generally declared in its permit and should be taken into account. Processes which cannot be subdivided, but which generate a very low contribution to the overall revenue may be neglected.

> Guidance: A block producer may manufacture two types of blocks on two separate lines. In this instance, it should be possible to avoid allocation by considering the actual inputs and outputs of each line, for example by using sub-meters to review the energy consumption, product mix to review the inputs, waste and production records to review the outputs. If the lines are not sub-metered, it is possible to allocate the energy used in the two lines using physical allocation, by considering the energy demand of the two lines, and the hours of operation. An aim of any physical allocation system should be to result in the impact that would occur if only one product was produced.
> In the case of coal-fired electricity and fuel ash, it is not possible to produce one without the other, and therefore this can be considered as a joint co-production process. To decide if any impacts need to be allocated to the fuel ash, then a co-product allocation procedure needs to be followed.
> Contributions to the overall revenue of the order of 1% or less are regarded as very low.

Joint co-product allocation shall be allocated as follows:

- *Allocation shall be based on physical properties (e. g. mass, volume) when the difference in revenue from the co-products is low;*
- *In all other cases allocation shall be based on economic values;*
- *Material flows carrying specific inherent properties, e. g. energy content, elementary composition (e. g. biogenic carbon content), shall always be allocated reflecting the physical flows, irrespective of the allocation chosen for the process.*

Where economic allocation is used to assess the impact of input materials being used within the product system, outputs from other systems which are not sold but given away or which the producer pays to dispose of, do not attract any burden from the production process to pass on to a further use. Outputs from manufacturing processes which are wastes but which reach the end-of-waste state where they have economic value, will attract a burden from their production system to pass on to a further use. Burdens for the further use will only be attracted from processes occurring once the end-of-waste state has been reached and will also be included in the system process using the secondary material.

Economic allocation for outputs of the studied product system will only be required where the outputs have reached the end-of-waste status and are sold. Where the revenue from such sales is very small, it can be ignored if required for simplicity of modelling.

Guidance: Where waste outputs leave the product system and enter another product system still as a waste at no cost to the next system, or at a cost to the waste producer, then no impacts from the waste producing system can be allocated to the next product system. In this instance, the next system is acting as a waste processing, waste recovery or waste disposal process.

Guidance Example: In many instances however it is not possible to produce only one of the products. For example, a quarry may produce both aggregate and dust, or stone blocks and aggregate and dust; or a BOF steelworks produces both steel and slags. In these instances, it is not possible to prevent the production of the unintended products, so allocation allows the impacts of producing all products to be split between the various products.

Guidance: Comparison of revenue from products should be undertaken using the market value per common unit of production (e. g. kg or MJ), based on normal pricing units and long term averages (3 or 5 years) if prices fluctuate. A difference in revenue of more than 25% is regarded as high. Market value of co-products may change over time, but it is the relation between them over time that is relevant, not the actual values. The actual overall revenue may be sensitive, but it is the % of revenue which is required for allocation. Rather than data for the whole plant, it could be considered at the cost of 1 kWh electricity and the cost of the resulting fuel ash etc. It may be clear, for instance, if the fuel ash is not a waste, that its relative value is less than 1% of the cost of the electricity.

混凝土 EPD 评价采用的 PCR 示例

> Where co-products have incompatible units, e. g. mass and energy, then it is essential to use economic allocation.
>
> Wherever possible, market value data should be obtained via the supplier of the waste or secondary material or a relevant local trade association. If this data is not provided, in many situations an estimate can be made of the values of the different co-products, goods and services, including waste processing services if relative values cannot be obtained from the suppliers. Market prices are the basic data to be used. If markets prices are not known there are several reliable sources on many product prices, including historical prices and expected prices in terms of futures. The web is a unique source of price data. Hundreds of web sites on most commonly traded products are available now. The relevant search term for market price is "fob", free on board, at the location of the supplier, without insurance or transport. The other price type is 'cif', stating a specific place of deliverance like "cif Chicago" for the price of slag as delivered in Chicago.
>
> Sources of relevant and appropriate data will vary depending on material and location, and should be considered on a case specific basis.

Physical properties such as sequestered carbon and calorific value remain a property of the material and cannot be allocated.

> Guidance: For example, the energy used to process logs into sawn lumber, sawdust and bark can be allocated between the 3 product flows, but the sequestered carbon within and calorific value of each co-product remains a physical property of each product flow.

The appropriate allocation procedure considered within this PCR for the following co-products or by-products which can be used within concrete production are provided below:

For any co-product or by-product or waste which is transported and used at no cost to the concrete producer, no allocation is required. The material is free of impact at the concrete factory gate.

> Guidance: In effect, if a co-product has no value then economic allocation will allocate no impact to it-we can therefore say that no allocation is required and the material is free of impact at the concrete factory gate.

For any co-product or by-product where the concrete producer is only paying for transport: no allocation of its production process is necessary; transport impacts are allocated to the concrete production.

Outside of Europe, agreements between industries on the allocation of impacts to co-products or wastes may be well established. In these cases, this PCR will respect these agreements though the Project Verification Report should provide information on the agreement.

> Guidance Example: There is already an agreement between the steel and cement industries regarding the allocation of impact to slag in Japan-in this instance no impact from steel production is transferred to the slag based on this agreement. Within Japan therefore, according to this PCR, slag will have no impact from the steel production process, irrespective of its waste status or value.

For any co-products or by-products where the concrete producer is paying for the raw material-see the examples below showing how allocation should be considered.

Granulated blast-furnace slag and air-cooled slag when they have reached the end of waste state-by economic value from pig iron produced in the blast-furnace. If not available, the ratio of the price of liquid slag to pig iron can be estimated by the ratio of the price of granulated slag versus steel billet.

- Zinc (or ISF) slag when it has reached the end of waste state-by economic value from zinc production
- Outputs from coal fired electricity power stations (fly ash, furnace bottom ash and cenospheres where they have reached the end-of-waste state)-by economic value from coal fired electricity production
- Incinerator ash which has reached the end-of-waste state-by economic value from waste treatment and incineration processes
- Foundry sand which has reached the end of waste state-by economic value from casting production
- Silica fume which has reached the end of waste state-by economic value from ferrosilicon or elemental silicon production
- China clay stent and sand which has reached the end of waste state-by economic value from china clay production
- Slate aggregate which has reached the end of waste state-by economic value from slate quarrying
- Oil shale which has reached the end of waste state-by economic value from oil

> Guidance Note that in some locations, these co-or by-products may not have reached the end-of-waste state or may have reached the end-of-waste state but be available at no cost to the concrete producer, in which case no allocation will be required (if the co-product has no value then no impact would be allocated on an economic basis).

混凝土 EPD 评价采用的 PCR 示例

> Guidance Example: Fly ash is a by-product from coal fired electricity production and may be sold to 'ash managers' who process and sell it to concrete producers. If the ash is sold to ash managers, then the value at the first point of sale is taken, and the impacts of the processing they undertake are within the concrete system boundary. If the fly ash is processed by the electricity producer, then sold directly to the concrete industry, then this cost would be taken for economic allocation. If the fly ash has no value and is given to the concrete industry, then no impact from the electricity production process is allocated to the concrete industry. The transport impacts from the transport of the "free" fly ash will be taken by whichever system pays for the transport.

2.3.5.3. Treatment of co-products produced during the product stage

Co-products or by-products which have never been waste are allocated as co-products at the gate of the product stage process which has produced them, using allocation procedures set out in 2.3.5.22.3.5.3.

Wastes from any production process in stages A1-A3 and the resulting impacts associated with their transport, treatment and processing are considered within the system boundary of the process producing the waste, until they reach the end-of-waste state. At this point, if they have an economic value then they can be considered as co-products of the process producing the waste and impact from that system can be allocated to them.

2.3.5.4. Treatment of co-products or by-products used during the product stage

Co-products or by-products from other products systems used during the product system bring impacts allocated from their previous system, using the allocation procedures set out in 2.3.5.5, plus any impacts arising within the system boundary. The system boundary for co-products or by-products which have never been wastes is set at the gate of the process that produced them. The system boundary for co-products or by-products which have been wastes is set at the point they reach the end-of-waste state (see 2.3.4.5).

2.3.5.5. Treatment of materials recovered from previous use (post-consumer waste)

No impacts are allocated over the system boundary from previous use for post-consumer material that is recycled or reused. This is the same for concrete after use at the system boundary-no impacts will be allocated over to any subsequent recycling. Impacts associated with any waste treatment or processes before the post-consumer material reaches the end-of-waste state are attributed to the system producing the post-consumer waste. All impacts occurring after the end-of-waste state is reached are attributed to the system using the post-consumer waste. For waste which has not reached the end-of-waste state, see 2.3.5.6.

Examples of post-consumer materials used in concrete and its supply chain which have reached the end-of-waste state are: glass cullet, recycled aggregate, recycled concrete aggregate. These materials will be free of impact from their first use. Impacts will only come from any processes after they reach the end-of-waste state.

2.3.5.6. Energy recovery of waste from other systems

Some wastes have not reached the end-of waste state before they become inputs to the production process where their inherent energy is recovered. In this case, the production process is acting as a waste treatment process for the waste, and theoretically the impacts of this treatment process should be part of the system that produced the waste. However, because of concerns that

- the interpretation of end-of-waste varies from location to location,
- emissions from the energy recovery of waste cannot easily be separated from the emissions from the use of other fuels,
- consistent results are obtained, aligned to the WBCSD CSI carbon reporting guidelines and
- to maintain a conservative approach,

within the WBCSD CSI PCR, the impacts from the energy recovery of waste will be included within the system boundary, but any indicators, particularly Global Warming Potential, which can be separately calculated for the energy recovery of wastes used as alternative fuels, can be reported as a sub-total of the total indicator.

> Guidance Example: If waste oil is an input in the cement kiln, the CO_2 emissions from the waste oil can be separately identified, based on the carbon content of the waste oil, and the SO_2 emissions from the Sulphur content of the waste oil and both can be reported as a sub-total as below.

Impact Category	Unit	Impact per m³ concrete
Global warming potential, GWP (TOTAL) -Global warming potential from energy recovery of wastes	kg CO_2 equiv	240 (64kg CO_2 eq. (27%) from energy recovery form wastes used as alternative fuels)
Depletion potential of the stratospheric ozone layer, ODP;	kg CFC_{-11} equiv	0.003
Acidification potential of soil and water, AP (TOTAL) -Acidification potential of soil and water from energy recovery of wastes	kg SO_2 equiv	1.7 (0.31kg SO_2 eq. (18%) from energy recovery from wastes used as alternative fuels)

Examples of wastes which are used in this way might include whole waste tyres or hazardous wastes which have not reached the end-of-waste state when they are used-for example when they are burnt in the cement kiln.

Impacts associated with any waste treatment or processes before the post-consumer material reaches the end-of-waste state are attributed to the system producing the post-consumer waste. All impacts occurring after the end-of-waste state is reached are attributed to the system using the post-consumer waste.

> Guidance Examples: For the combustion of materials not considered to have reached the end-of-waste state, at a minimum, carbon dioxide emissions and any other emissions which can be accurately estimated for the burning of the waste should be attributed to the system producing the waste, not the system using the waste. For other emissions, co-product allocation based on the value of the waste treatment function could be considered, but revenue from this type of waste treatment is normally extremely low relative to other revenue from manufacturing, so these other emissions can be considered within the system boundary for the purposes of this PCR.

2.3.5.7. Treatment of biogenic carbon

Biogenic carbon is the carbon taken up and incorporated within biomass during growth (also known as sequestered carbon) and which can be released during decomposition or combustion at the end of life. Biogenic carbon uptake is considered as the net uptake of CO_2 within the system boundary over the 100 years before the biomass is harvested (see 2.3.4.2 Fel! Hittar inte referenskäla.) and can therefore be calculated from the carbon content of the biomass. Biogenic carbon release is considered from the combustion of biomass at the end of life, and from its behaviour in landfill over 100 years from deposition (see 2.3.4.2).

Sequestered carbon is considered as an inherent physical property of the biomass and cannot be allocated by other mechanisms (see 2.3.5.2 Fel! Hittar inte referenskäla.).

Land use change must be considered where relevant (see 2.6.2).

> Guidance: For biomass used as a fuel, the biogenic carbon released during combustion will match the biogenic carbon uptake during growth. This biogenic carbon can either be modelled on both sides (as an uptake and emission which will balance each other) or the biogenic uptake and emission can both be ignored. In either instance, there may be other carbon emissions from extraction, processing and transport which will need to be considered and which can be allocated.
> Unsustainably harvested timber would be associated with land use change (release of soil based carbon) which must be considered in the assessment. Sustainable harvested timber is not associated with land use change (see 2.6.2).

2.3.5.8. Criteria for the exclusion of inputs and outputs

Criteria for the exclusion of inputs and outputs (cut-off rules) in the LCA and information modules and any additional information are intended to support an efficient calculation pro-

cedure. They shall not be applied in order to hide data. Any use of cut-off criteria for the exclusion of inputs and outputs shall be documented.

The following procedure shall be followed for the exclusion of inputs and outputs:

- All inputs and outputs to a (unit) process shall be included in the calculation, for which data are available, unless otherwise stated in this PCR.

Data gaps may be filled by conservative assumptions with average or generic data. Any assumptions for such choices shall be documented;

- In case of insufficient input data or data gaps for a unit process, the cut-off criteria shall be 1% of renewable and non-renewable primary energy usage and 1% of the total mass input of that unit process. The total of neglected input flows for the stages "cradle through construction" shall be a maximum of 5% of energy usage and mass. Conservative assumptions in combination with plausibility considerations and expert judgement can be used to demonstrate compliance with these criteria;

- Particular care should be taken to include material and energy flows known to have the potential to use significant resources or cause significant emissions into air and water or soil related to the environmental indicators of this PCR.

Conservative assumptions in combination with plausibility considerations and expert judgment can be used to demonstrate compliance with these criteria.

2.3.5.9. Capital equipment and infrastructure

The contribution of capital equipment and infrastructure is not normally considered in LCA as the share of impact of this equipment per unit of production almost always falls below the study's cut-off criteria. It is not included here unless it is significant, in accordance with the data cut-off rule. Maintenance of equipment is likewise not included in the LCA with the exception of regularly consumed items such as mould oil which are included in the inventory based on the average amount used per year if they exceed the thresholds defined in the cut-off rules.

> Guidance Examples: The machines and equipment that are used in factories, and the factories themselves have an impact in their manufacture and disposal. However, this impact is generally very small in comparison to the impact of the product that they produce over their life time. The operation of machines and equipment, and operation of factories (heating and lighting) is included within the scope.

2.3.5.10. Head office activity, marketing, sales etc

All energy used in factories and factory support offices is included. The impacts associated with company management, marketing and sales activities which may be located either within factory sites or at other locations may be excluded from the assessment.

2.3.6 UNITS

SI units shall be used. Basic units include: metre (m), kilogram (kg), and moles (mol) (an amount of chemical substance). Derived SI units include kg/m^3 or MJ ($kg \cdot m^2/s^2$) for example. For very small or very large quantities, SI prefixes such as "kilo" or "milli" can be used, eg. kilometre for transport distances or millimetre for product thicknesses.

All resources flows with the exception of energy and water consumption are expressed in kg.

- Resources used for energy input (primary energy) are expressed as kWh or MJ, including renewable energy sources e. g. hydropower, wind power;
- Water consumption which is reported in m^3.

Exceptions are:

- Temperature, which is expressed in degrees Celsius;
- Time, which is expressed in practical units depending on the assessment scale: minutes, hours, days, years.

2.4 INVENTORY ANALYSIS

2.4.1 COLLECTING DATA

Data collection shall follow the guidance provided in EN ISO 14044: 2006, 4.3.2, see below.

> *EN ISO 14044: 2006, 4.3.2 The qualitative and quantitative data for inclusion in the inventory shall be collected for each unit process that is included within the system boundary. The collected data, whether measured, calculated or estimated, are utilized to quantify the inputs and outputs of a unit process. When data have been collected from public sources, the source shall be referenced. For those data that may be significant for the conclusions of the study, details about the relevant data collection process, the time when data have been collected, and further information about data quality indicators shall be referenced. If such data do not meet the data quality requirements, this shall be stated. To decrease the risk of misunderstandings (e. g. resulting in double counting when validating or reusing the data collected), a description of each unit process shall be recorded. Since data collection may span several reporting locations and published references, measures should be taken to reach uniform and consistent understanding of the product systems to be modelled.*
>
> *EN ISO 14044: 2006 4.3.2.2 These measures should include the following:*
>
> - *drawing unspecific process flow diagrams that outline all the unit processes to be modelled, including their interrelationships;*
> - *describing each unit process in detail with respect to factors influencing*

inputs and outputs;
- listing of flows and relevant data for operating conditions associated with each unit process;
- developing a list that specifies the units used;
- describing the data collection and calculation techniques needed for all data;
- providing instructions to document clearly any special cases, irregularities or other items associated with the data provided.

2.4.2 CALCULATION PROCEDURES

The calculation procedures described in EN ISO 14044 shall apply. The same calculation procedures shall be applied consistently throughout the study.

Guidance Example: Over a year, a concrete plant may, amongst other inputs, use 12000kWh of grid electricity to make all its concrete (20,000m^3 total production), and 3000 tonnes of aggregate to make 2000m^3 of a specific type of concrete, eg GEN 1. The different concretes are produced in a similar manner and therefore the amount of electricity will be the same for a m^3 of each. The amount of aggregate will vary for each type of concrete and therefore needs to be calculated for the specific volume of concrete produced, on the basis of the mix design. The reference unit for both the typical and specific concrete produced will be 1m^3. The inputs need to be calculated on the basis of the reference unit, so the plant will use 12000/20000 = 0.6kWh = 2.16MJ of grid electricity per m^3 of any concrete, and 3000/2000 = 1.5tonnes = 1500kg of aggregate per m^3 of GEN 1 concrete. Once this initial input inventory has been calculated, the inventory needs to be linked to LCI datasets for the input materials. The LCI datasets are provided for a relevant reference unit, for example from the ELCD (http://lca.jrc.ec.europa.eu/lcainfohub/datasetList.vm) electricity data is provided per 3.6MJ and crushed stone per kg; if units do not correspond data must be converted. LCI data for 3.6MJ of grid electricity therefore needs to be multiplied by the use of electricity (in MJ/3.6) for 1m^3 of concrete (the reference unit). So the dataset for 1 MJ electricity will be multiplied by 2.16/3.6 = 0.6 and the dataset for 1 kg of aggregate by 1500.

For the use of 3.6 MJ French Grid Electricity, 0.159MJ of energy from Crude Oil are used as a resource and 0.134kg Carbon Dioxide are emitted amongst other inventory, so 2.16MJ will use 0.0954MJ crude oil and produce 0.0804kg Carbon Dioxide.

For crushed stone, cradle to gate production of 1kg uses 1.103kg calcium carbonate as a resource from nature, and produces 0.013kg Carbon Dioxide amongst other inventory, so 1500kg will use 1654.5kg calcium carbonate as a resource and produces 19.59kg Carbon Dioxide.

混凝土 EPD 评价采用的 PCR 示例

Once all inputs and outputs have been linked with all the relevant LCI flows, the inventory can be aggregated to calculate the total resources and emissions from the production of $1m^3$ of typical concrete or $1m^3$ of the specific concrete, eg GEN 1.

The next stage of the LCA calculation is to assess which resources and emissions are linked to which environmental indicators, a process known as "classification". For example carbon dioxide causes Global Warming impacts. For each impact and flow, the degree to which a reference flow causes the impact is used to "characterise" the impacts, with characterisation factors being published for various impact categories. For example, the IPCC publish Global Warming Potentials which are used to characterise the Global Warming impact of greenhouse gases-Carbon dioxide has a factor of 1, methane of 25 with the 100 year timeframe used within this PCR.

Once all the resource and emissions flows have been classified and characterised for each indicator category used in the EPD, the results for each item of inventory are added to give the total for each indicator for $1m^3$ of typical concrete.

Note: Most LCA software will include the classification and characterisation data required by this PCR and undertake many of these processes automatically.

When transforming the inputs and outputs of combustible material into inputs and outputs of energy, the net calorific value of fuels shall be applied according to scientifically based and accepted values specific to the combustible material.

Guidance: Net calorific value (NCV) or Lower Heating Value (LHV) of fuels are routinely measured at plant level. It is important to note that the applied heating value always has to match the status of the fuel, especially with respect to the correct moisture content during its weighing (e.g. raw coal or dried coal). Normally the lower heating value is determined from a dried sample. Subsequently a moisture correction has to be applied to the result, correcting the mass reference from the dried sample back to the original moisture content of the fuel as it is consumed or weighed.

For the conversion of higher heating values (HHV or gross calorific value GCV) to LHV the equation defined in the 2006 IPCC Guidelines4 (Vol. II, Section 1.4.1.2, Box 1.1) can be applied. [Taken from WBCSD CSI "The Cement CO_2 and Energy Protocol".

Guidance Examples: When determining the elementary flows associated with production, the actual production mix should be used whenever possible, in order to reflect the various types of resources that are consumed. As an example, for the production and delivery of electricity, account shall be taken of the electricity mix, the efficiencies of fuel combustion, conversion, transmission and distribution losses.

All calculation procedures shall be explicitly documented in the Project Verification Report and the assumptions made shall be clearly stated and explained.

> EN ISO 14044: 2006, *Care should be taken when aggregating the inputs and outputs in the product system. The level of aggregation shall be consistent with the goal of the study. Data should only be aggregated if they are related to equivalent substances and to similar environmental impacts. If more detailed aggregation rules are required, they should be explained in the goal and scope definition phase of the study or should be left to a subsequent impact assessment phase.*

Guidance: Aggregating data occurs when, for example, the products of a number of concrete plants are considered within a single EPD. In this case, the data can either be aggregated vertically, by calculating the impacts of each site separately using their own supply chain data and combining the data based on weighted production; or it can be aggregated horizontally, for example by calculating the typical impacts of cement plants used by the concrete plants, and then using this data within the an assessment of all the concrete plants. Horizontal aggregation may be less accurate if, for example, the distribution of cement plants used does not match the production of the cement plants so that the weighted average is incorrect, but where national or trade association study is being undertaken should produce similar results.

2.4.3 SELECTION OF DATA

As a general rule, specific data derived from specific production processes or average data derived from specific production processes shall be the first choice as a basis for calculating an EPD. In addition the following rules apply:

- An EPD describing an average product shall be calculated using representative mean average data of the products declared by the EPD;
- An EPD describing a specific product shall be calculated using specific data for at least the processes the producer of the specific product has influence over. Generic data may be used for the processes the producer cannot influence e.g. processes dealing with the production of input commodities, e.g. raw material extraction or electricity generation, often referred to as upstream data (see Table 1);
- The additional technical information for the development of scenarios of the building's life cycle stages shall be specific or specific average information, when an average product is declared;
- Documentation of technological, geographical and time related representativeness for generic data shall be provided in the Project Verification Report.

Application of generic and specific data   Table 1

MODULES	ModuleA1-A3	
	Production of commodities, raw materials	Product manufacture
Process type	Upstream processes	Processes the manufacturer has influence over
Data type	Generic data	Manufacturer's average or specific data

> Guidance: Generic data is publicly available and may be average or specific. Normally it is used to describe upstream and downstream processes. See CEN/TR 15941, Sustainability of construction works—Environmental product declarations—Methodology for selection and use of generic data.

2.4.4 DATA QUALITY REQUIREMENTS

The quality of the data used to calculate an EPD shall be addressed in the Project Verification Report (see Clause 8 and EN ISO 14044: 2006, 4.2.3.6). In addition the following specific requirements apply:

- Data shall be as current as possible. Data sets used for calculations shall have been updated within the last 10 years for generic data and within the last 5 years for producer specific data;
- Data sets shall be based on 1 year averaged data; any deviations shall be justified;
- The time period over which inputs to and outputs from the system shall be accounted for is 100 years from the year for which the data set is deemed representative.

> Guidance: For waste disposal processes such as landfill, emissions to nature shall be measured for 100 years from the point of disposal.
>
> Guidance: For virgin biomass based products, the system will include all inputs and outputs from the product system for 100 years before the point of harvest (e.g. for Timber) or the normal agricultural crop cycle if this is shorter.

- The technological coverage shall reflect the physical reality for the declared product or product group;
- Generic data: Guidance for the selection and use of generic data is provided in CEN/TR 15941. Generic data shall be checked for plausibility;
- Data sets shall be complete according to the system boundary within the limits set by the criteria for the exclusion of inputs and outputs, (see 2.3.5.3).

The use of upstream data, which does not respect the allocation principles described in this PCR shall be clearly stated and justified in the Project Verification Report. These data shall be in line with EN ISO 14044 allocation rules.

> Guidance: A check on data validity shall be conducted during the process of data collection to confirm and provide evidence that the data quality requirements for the intended application have been fulfilled. Validation may involve establishing, for example, mass balances, energy balances and/or comparative analyses of release factors. As each unit process obeys the laws of conservation of mass and energy, mass and energy balances provide a useful check on the validity of a unit process description. Obvious anomalies in the data resulting from such validation procedures require alternative data that do comply with the data selection rules.

2.5 TYPES OF EPD

2.5.1 EPD COVERING MORE THAN ONE PRODUCT

In cases where several similar products are produced by a site or company, the PCR offers the possibility for similar products to be grouped as an average product in the same EPD. In this case a mass weighted average of production should be used to calculate the average for the product group.

In the case that the difference of the environmental impacts between the products is higher than 10%, information on the range of variation should be provided as per section 3.1 j).

> Guidance Example: Manufacturer X may produce a range of products which are very similar-for example solid and hollow blocks of different sizes. They can provide an EPD with a single set of values to cover the range of products. In case the difference of the environmental between the products is higher than 10% the producer should provide information in the EPD to convey the range of impact across the product range-for example by giving information such as +10%-13% for each indicator, or by giving the standard deviation for each indicator. This range only applies to the variation between the different products, not to the uncertainty associated with any individual dataset.

2.5.2 SECTOR EPD

It is also possible to create a so-called Sector EPD which enables the possibility to present average data for a whole industrial branch in a well-defined geographical area.

Where a group of manufacturers are declaring performance using a single sector EPD, then a mass weighted average of production should be used to calculate the average for the product or product group.

Where the average for a product group is provided, information on the range of variation should be provided as per section 3.1 j).

> Guidance Example: A regional trade association may provide a sector EPD for the range of concrete products that are produced. In this case, the range can be covered by a single set of values, but information should be provided in the EPD to convey the range of impact across the product range-for example by giving information such as +10%-13% for each indicator, or by giving the standard deviation for each indicator. This range only applies to the variation between the different plants providing data, not to the uncertainty associated with any individual dataset.

2.6 OTHER ASPECTS

2.6.1 ELECTRICITY

Where electricity is generated on-site, then specific data for the inputs and emissions should be used. If electricity is sourced from the national grid, then the LCI associated with the national grid where the life cycle stage occurs shall be used. When a supplier of electricity can deliver a specific electricity product and guarantee that the electricity sale and the associated emissions are not double counted, the data for that electricity shall be used for the product studied. When the supplier of electricity does not provide specific data for the specific electricity product, then use of the national grid should be taken.

If specific life cycle data on a process within the energy supply system are difficult to access, data from recognized databases may be used.

The treatment of electricity should be documented.

> Guidance: If electricity is generated on site, then data for inputs and emissions should be collected as part of the overall site data collection.
> Although primary data should be used for the emissions from the power generated, the installations with onsite-power generation (other than Waste Heat Recovery), should be considered as indirect emission (included in module A1) to allow comparison.
> If supplier specific data is used, any additional data, such as the inclusion of upstream processes, e.g. associated with any fuel extraction, processing and transport, or other default emissions, must be considered to ensure the data used is cradle to gate data covering all relevant emissions and resources, comparable in scope to national data sets.

Where a country does not have a single national electricity grid but has several unconnected grids, the relevant grid from which the power is obtained should be used.

Where a national grid is part of a larger regional grid, national data, taking into account imports and exports (if significant) to the regional grid should be used.

> Guidance: Regarding double-counting, generator-specific emission factors for electricity used in a process could be used when:
> - the process used the electricity (or used an equivalent amount of electricity of the same type to that generated), and another process did not claim the generator-specific emission factors for that electricity; and
> - the generator-specific electricity production does not influence the emission factors of any other process or organization.
>
> In some countries, parts of the electricity from renewable energy sources might already be sold/exported as "green" electricity, and should thus be excluded from the mix to avoid double counting.
>
> Some "green certificates" are sold without coupling to the electricity, which might lead to double counting.
>
> Guidance Examples: If a manufacturer had PV on their roof, if they use all the output in their plant and never export any to the grid or other users then the LCA may use a model of LCA for PV electricity for the quantity of electricity obtained from these PV panels. If, however, they sell the PV electricity to the grid or other users, then they cannot benefit from this "sold" electricity. Grid sourced green electricity can only be included if it can be demonstrated that they are the only company claiming this benefit (i.e. that the grid is not greener as a result, or other users are also "buying" the same electricity through other types of tariffs (possible in the UK).

2.6.2 LAND USE CHANGE

When significant, the GHG emissions and removals occurring as a result of direct land use change (dLUC) shall be assessed in accordance with internationally recognized methods such as the Intergovernmental Panel on Climate Change (IPCC) Guidelines for National Greenhouse Gas Inventories and included in the life cycle inventory. Land use change GHG emissions and removals shall be documented separately in the Project Verification Report. If site-specific data are applied, they should be transparently documented in the Project Verification Report.

Guidance: If land use change occurs at the start of a long term process, for example quarrying activity over many years, then the impact should be considered over the total production of the quarry and included if significant.

Guidance Examples: Land use change may occasionally occur when forestry is cleared for quarrying, or when timber is illegally harvested as a fuel. Illegal forest clearance or use of timber which is harvested illegally must take into account the land use change impact associated with such a change of use. This will mean that additional impacts associated with land use change will be included, alongside the consideration of biogenic carbon uptake and biogenic carbon emission at end of life if timber is used as fuel.

2.7 IMPACT ASSESSMENT

2.7.1 ENVIRONMENTAL IMPACT CATEGORIES

The impact assessment shall be carried out for the following environmental impact categories:
- *global warming*;
- *ozone depletion*;
- *acidification of land and water*;
- *eutrophication*;
- *photochemical ozone creation*;
- *depletion of abiotic resources* (*elements*);
- *depletion of abiotic resources* (*fossil*).

These should be calculated using characterisation factors recommended in regionally accepted impact assessment methods. In Europe, the latest CML baseline indicators should be used. The US Environment Protection Agency (EPA) recommends methods used in TRACI while in Australia, the Building Products Innovation Council (BPIC) has published applicable impact assessment methods for that region.

> Guidance:
> In Europe, the latest baseline characterisation factors shall be taken from CML (Institute of Environmental Sciences Faculty of Science University of Leiden, Netherlands). The characterisation factors for ADP-fossil fuels are the net calorific values at the point of extraction of the fossil fuels. CML characterisation factors can be downloaded at http://cml.leiden.edu/software/data-cmlia.html or are available in many commercial LCA softwares.
> In the US, TRACI (US EPA Tool for the Reduction and Assessment of Chemical and other environmental Impacts) uses specific methodology and units for the assessment of acidification (mole H^+ equiv), Eutrophication (k N equiv) and Smog Creation Potential (kg O_3 equiv). See http://www.epa.gov/nrmrl/std/sab/traci/
> In Australia, the BPIC/ICIP project has recommended indicators, and some indicator methodologies for the life cycle assessment of construction products in Australia. The recommended indicators are described in http://www.bpic.asn.au/literature79925/ Life Cycle Impact Assessment-Part 1 Classification and Characterisation.

In addition to these, other regionally accepted impact categories may be reported if required.

2.7.2 USE OF NET FRESH WATER

2.7.2.1. Water inventory

Data related to water which represent elementary flows and technical flows may be directly collected from unit processes or derived from LCA data for upstream processes, e.g. electricity or waste for further processing. Generally categories of data related to each elementary flow and technical flow for inputs and outputs should include:

a) Quantities of water withdrawn (used) and discharged:

-mass, or volume (e.g. water entering and leaving the unit process);

b) Water withdrawal of fresh water and non-fresh water:

Source of water withdrawn (used) for elementary flows (resource types), e.g.:

-rainwater;

-surface water (water from rivers, lakes, ponds…);

-groundwater (renewable, excluding fossil water), and

-fossil water (non-renewable groundwater);

-sea water;

-brackish water;

Source of water withdrawn (and used) for technical flows:

-municipal water supply system

c) Water discharge of fresh water and non-fresh water:

Fresh water discharge by receiving body: Ocean, surface, well, of-site water treatment

Non-fresh water discharge by receiving body: Ocean, surface, well, of-site water treatment

d) Emission of water: evaporation water, evapotranspiration water

e) water quality parameters:

-e.g. chemical, physical (e.g. thermal), and biological characteristics;

f) Forms of water use:

-different forms of water consumption, e.g. evaporation, evapotranspiration, product integration, discharge into different drainage basins or the sea;

g) Geographical location of water withdrawal and return;

-information on the physical location of water withdrawal and return (as site-specific as possible) or assignment of the physical locations to a category derived from an appropriate classification of drainage basins or regions;

The water inventory shall include inputs and outputs from each unit process being part of the system to be studied. Any discrepancies in the inventory balance shall be explained.

> Guidance: Information on location is required so that in the future when robust methods are in place, it will be possible to determine any related environmental condition indicator (e. g. water stress, local level of social development, etc.) of the area where the water use takes place.
>
> At present, only very few robust methods to develop impact assessment for water consumption are available which relate to water consumption on a watershed level (drainage basin level). Methods considering water consumption on the level of water bodies (e. g. withdrawal of groundwater and discharge of this groundwater to surface water) are only in development. Therefore, the EPD only provides aggregated data on net fresh water consumption. It is recognized that in future this aspect of methodology will need further development and for this reason, data on the geographical location should be recorded. A small selection of local/regional water scarcity assessment measures is already available and these could be used to report on the environmental impacts of water consumption as additional information.
>
> Tap water and purified waste water are not elementary flows but intermediate flows from a process within the technosphere ("technical flows", e. g. from a water treatment plant). Use of tap water or purified water will bring burdens from the treatment and distribution process.
>
> Where electricity is generated on-site, then specific data for the inputs and emissions should be used. Only in the case of the Waste Heat Recovery (WHR), the water consumption would be considered as being integrated within the process.

2.7.2.2. Net fresh water consumption

Net fresh water consumption describes the amount of fresh water withdrawn for which release back to the source of origin (e. g. drainage basin or sea). does not occur. Net fresh water consumption is calculated as the difference between the sum of water withdrawals of fresh water and the sum of water discharges stemming from fresh water, including amounts of fresh water lost by evaporation, evapotranspiration, product integration and release to a different source of origin (e. g. different drainage basin or sea). Fresh water consumption includes the following water resource types reported for each unit process:

- rainwater;
- surface water;
- groundwater (excluding fossil water), and
- fossil water.

The net fresh water consumption for a unit process is the sum of the calculated water consumption for four fresh water types. The net fresh water consumption for a group of processes or a product is the sum of the net fresh water consumption for all the upstream unit processes.

Guidance Examples:

Quarry dewatering: If quarry water is just pumped from the quarry and placed in the river then this does not need to be accounted for as water use or consumption. Nevertheless, where evaporation may occur (particularly in hot climates), the share of evaporative loss should be considered for quarry water from ground water resources temporarily stored in artificial open ponds/lagoons or water storage facilities. Consequently, this evaporated quarry water should actually be reported as water consumption.

Rain and storm water: Rain and storm water used on the site should be reported (on input and output side) to facilitate sound water balances.

Recycled water: Recycled water stored in an open pool or a settlement lagoon may be subject to evaporation (particularly in hot climates). Thus, the evaporative loss of stored recycled water should be considered in the water balance as water consumption. On the input side, this loss is accounted for by additional fresh water withdrawals.

Guidance: Estimated evaporation can be used when consumption from this route is considered to be significant and accurate quantities are not available. The following document may be helpful in providing calculation of estimates: US EPA "Risk Management Program Guidance for Site Consequence Analysis" which can be downloaded from http://www.epa.gov/oem/docs/chem/oca-chps.pdf.

3 CONTENT OF THE EPD

3.1 DECLARATION OF GENERAL INFORMATION

The following items of general information are required and shall be declared in an EPD.

a) the name and address of the manufacturer (s);
b) the description of the construction product's use and the declared unit of the construction product to which the data relates;
c) construction product identification by name (including any product code) and optionally, a simple visual representation of the construction product to which the data relates;
d) a description of the main product components and or materials;

> Guidance: This description is intended to enable the user of the EPD to understand the composition of the product represented in the EPD as delivered and also support safe and effective installation, use and disposal of the product.

e) a content declaration of the product covering relevant materials and substances. The gross weight of material shall be declared in the EPD at a minimum of 99%.

The declaration of material content of the product shall list as a minimum substances contained in the product that are listed in the "Candidate List of Substances of Very High Concern (SVHC) for authorisation" when their content exceeds 0.1 weight-% of the product. SVHC are listed by European Chemicals Agency and includes the Candidate List of SVHC (see http://echa.europa.eu/chem_data/authorisation_process/candidate_list_table_en.asp).

An optional detailed list of the product's substances, including CAS number, environmental class and health class, may be included in the product content declaration. It is also recommended to include substances' functions in the product (e.g., pigment, preservative, etc.).

> Guidance: The source location of any safety data sheet can be provided.

f) name of the programme used and the programme operator's name and address and, if relevant logo and website;

> Guidance: The EPD program or programs where this PCR will be registered have not yet been decided.

g) the date the declaration was issued and the 5 year period of validity;
h) A statement that the EPD only covers the Cradle to Gate stage, or the Cradle to Gate plus construction stage, because other stages are very dependent on particular sce-

narios and are better developed for specific building or construction works.
i) a statement that EPD of construction products may not be comparable if they do not comply with the requirements of comparability set in EN 15804;
j) In the case where an EPD is declared as an average environmental performance for a number of products, a statement to that effect shall be included in the declaration together with a description of the range / variability of the LCIA results if significant;
k) the site (s), manufacturer or group of manufacturers or those representing them for whom the EPD is representative;
l) information on where explanatory material may be obtained.

Guidance on safe and effective installation, use and disposal of the product can be supplied.

In addition to the above-mentioned general information, Table 2 shall be completed and reproduced in the EPD.

Demonstration of verification   Table 2

The PCR UN CPC 375 serves as the PCR for this EPD
Independent verification of the declaration, according to ISO 14025:2006 ☐ internal   ☐ external
Independent Verifier: <Name and Organisation of the Independent verifier>

Independent verifiers, whether internal or external to the organization, shall not have been involved in the execution of the LCA or the development of the declaration, and shall not have conflicts of interests resulting from their position in the organisation.

Guidance: Independent verification is essential for any form of external communication.

3.2 DECLARATION OF ENVIRONMENTAL PARAMETERS DERIVED FROM LCA

3.2.1 GENERAL

To illustrate the product system studied, the EPD shall contain a simple flow diagram of the processes included in the

An example of a process flow diagram for a cement kiln taken from WBCSD CSI Cement CO_2 and Energy Protocol is included below.

混凝土 EPD 评价采用的 PCR 示例

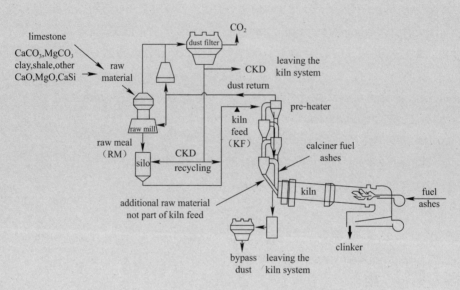

Figure 3-1: Example of a cement process flow diagram.

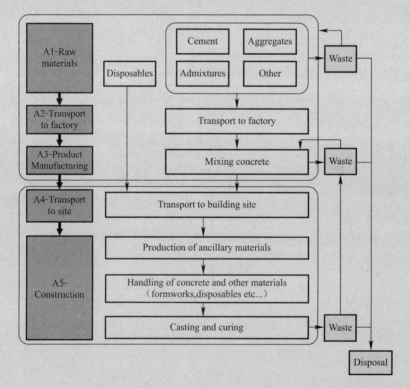

Figure 3-2: Example of a ready-mixed concrete process flow diagram

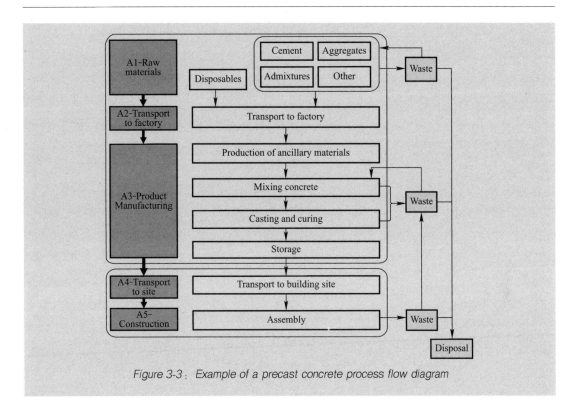

Figure 3-3: Example of a precast concrete process flow diagram

3.2.2 PARAMETERS DESCRIBING ENVIRONMENTAL IMPACTS

The following information on environmental impacts is expressed with the impact category parameters of LCIA using characterisation factors. These predetermined parameters are required and shall be included in the EPD as follows:

Parameters describing environmental impacts　　　　　　　Table 3

Impact Category	Parameter	Parameter unit expressed per functional/declared unit
Global Warming	Global warming potential, GWP (100years). Characterisation Factors: International Panel for Climate Change 4th Assessment Report, 2007.	kg CO_2 equiv
Ozone Depletion	Depletion potential of the stratospheric ozone layer, ODP;	kg CFC 11 equiv
Acidification for soil and water	Acidification potential of soil and water, AP;	mol H+ equiv /kg SO_2 equiv
Eutrophication	Eutrophication potential, EP;	mol N equiv /kg PO4 equiv
Photochemical ozone creation	Formation potential of tropospheric ozone, POCP;	kg NMVOC equiv /kg Ethene equiv
Depletion of abiotic resources-elements	Abiotic depletion potential (ADP-elements) for non fossil resources[a]	kg Sb equiv
Depletion of abiotic resources-fossil fuels	Abiotic depletion potential (ADP-fossil fuels) for fossil resources[a]	MJ, net calorific value
[a] The abiotic depletion potential is calculated and declared in two different indicators: • ADP-elements: include all non renewable, abiotic material resources (i.e. excepting fossil resources). • ADP-fossil fuels include all fossil resources.		

Guidance: Regionally accepted publications carry details of these indicators and source documents for characterisation factors. Brief descriptions, for the specific case of Europe, are provided below.

Abiotic Depletion Potential (ADP) (elements): impact from the depletion of scarce non-renewable resources such as metals, expressed in comparison to the element antimony.

Characterisation factors based on the ultimate reserves are provided in Oers, L. F. C. M., van & Koning, A., de & Guinée, J. B. & Huppes, G., 2002. Abiotic resource depletion in LCA: improving characterisation factors for abiotic depletion as recommended in the new Dutch LCA Handbook. Delft: Ministry of Transport, Public Works and Water Management. Baseline factors using the ultimate reserve have also been included in CML-IA which can be downloaded at http://cml.leiden.edu/software/data-cmlia.html.

Abiotic Depletion Potential (ADP) (fossil): impact from depletion of fossil fuel resources such as oil or natural gas, expressed using their net calorific value.

Characterisation factors are based on the net calorific value of the fossil fuel resource. Indicative factors have been included in CML-IA and can be downloaded at http://cml.leiden.edu/software/data-cmlia.html.

Global Warming Potential (GWP): global warming impact of greenhouse gases such as Carbon Dioxide (CO_2), measured using the equivalent CO_2 emission over a 100 year time horizon.

Characterisation factors are provided in Table 2.14 of the IPCC 4th Assessment Report which can be viewed at http://www.ipcc.ch/publications_and_data/ar4/wg1/en/ch2s2-10-2.html. These factors are also included in the CML-IA baseline indicators (http://cml.leiden.edu/software/data-cmlia.html).

Ozone Depletion Potential (ODP): relative impact that the product can cause to the stratospheric ozone layer, compared to an emission of trichlorodifluoromethane (CFC-11), using a 100 year timescale.

Characterisation factors are provided in 'Scientific Assessment of Ozone Depletion: 1998' World Meteorological Organization Global Ozone Research and Monitoring Project-Report No. 44. The executive summary of this report is located at http://www.esrl.noaa.gov/csd/assessments/ozone/1998/executive_summary.html but characterisation factors cannot be downloaded there. These factors are included in CML-IA baseline indicators (http://cml.leiden.edu/software/data-cmlia.html).

Acidification Potential (AP): increase of soil and water acidity that the product can cause, measured in kg of Sulphur Dioxide equivalent.

Characterisation factors are provided by Huijbregts, M., 1999b: Life cycle impact assessment of acidifying and eutrophying air pollutants. Calculation of equivalency factors with RAINS-LCA. Interfaculty Department of Environmental Science, Faculty of Environmental Science, University of Amsterdam, The Netherlands. These factors are included in CML-IA baseline indicators (http://cml.leiden.edu/software/data-cmlia.html).

Eutrophication Potential (EP): impact of nutrification by nitrogen and phosphorus to aquatic and terrestrial ecosystems, for example through algal blooms, disturbing the balance between species, measured in moles of Nitrogen equivalence.

Characterisation factors are provided by Huijbregts, M., 1999b: Life cycle impact assessment of acidifying and eutrophying air pollutants. Calculation of equivalency factors with RAINS-LCA. Interfaculty Department of Environmental Science, Faculty of Environmental Science, University of Amsterdam, The Netherlands. These factors are included in CML-IA baseline indicators (http://cml.leiden.edu/software/data-cmlia.html).

Photochemical Ozone Creation Potential (POCP): also known as summer smog, the impact from oxidizing of volatile compounds in the presence of nitrogen oxides (NOx) which frees ozone in the low atmosphere, measured relative to Ethene (C2H4).

Characterisation factors are provided Jenkin, M.E. & G.D. Hayman, 1999: Photochemical ozone creation potentials for oxygenated volatile organic compounds: sensitivity to variations in kinetic and mechanistic parameters. Atmospheric Environment 33: 1775-1293 and Derwent, R.G., M.E. Jenkin, S.M. Saunders & M.J. Pilling, 1998. Photochemical ozone creation potentials for organic compounds in Northwest Europe calculated with a master chemical mechanism. Atmosperic Environment, 32. p 2429-2441. These factors are included in CML-IA baseline indicators (http://cml.leiden.edu/software/data-cmlia.html).

Commercial LCA software also implement many of these life cycle impact assessment methodologies.

3.2.3 PARAMETERS DESCRIBING RESOURCE USE

The following environmental parameters apply data based on the LCI. They describe the use of renewable and non-renewable material resources, renewable and non-renewable primary energy and water. They are required and shall be included in the EPD as follows:

Parameters describing resource use   *Table 4*

Parameter	Parameter unit expressed per functional/declared unit
Use of renewable primary energy excluding renewable primary energy resources used as rawmaterials	MJ, net calorific value
Use of renewable primary energy resources used as raw materials	MJ, net calorific value

混凝土 EPD 评价采用的 PCR 示例

续表

Parameter	Parameter unit expressed per functional / declared unit
Total use of renewable primary energy resources (primary energy and primary energy resources used as raw materials)	MJ, net calorific value
Use of non-renewable primary energy excluding non-renewable primary energy resources used as raw materials	MJ, net calorific value
Use of non-renewable primary energy resources used as raw materials	MJ, net calorific value
Total use of non-renewable primary energy resources (primary energy and primary energy resources used as raw materials)	MJ, net calorific value
Use of secondary material	kg
Use of renewable secondary fuels	MJ, net calorific value
Use of non-renewable secondary fuels	MJ, net calorific value
Use of net fresh water	m^3

Guidance: In order to identify the input part of renewable/non-renewable primary energy used as an energy carrier and not used as raw materials, the parameter "use of renewable/non-renewable primary energy excluding renewable/non-renewable primary energy resources used as raw materials" is considered and can be calculated as the difference between the total input of primary energy and the input of energy resources used as raw materials.

Net fresh water has been considered equivalent to the concept of consumption of fresh water used in the developing ISO Standard 14046 for water footprinting. A description of the calculation of consumption of net fresh water based on this developing Standard is provided in 2.7.2.

It is also possible to report fresh water consumption at a more granular level covering levels of water stress and/or by different water type. Recommended water stress indicators for geographical areas are available for download from the website http://www.ifu.ethz.ch/ESD/data/.

Guidance: Brief descriptions of these indicators are provided below.

Primary energy (or embodied energy): total energy resources required to manufacture the product. Non-renewable energy derives from fossil fuels and uranium, and renewable energy from biomass, wind, solar or hydraulic sources.

Water consumption (or embodied water): water used in the production process, which can be a significant proportion of the total footprint. It does not take into account any aspect of geographical scarcity.

Where relevant, these indicators may also provide data on the impacts associated with the use of energy recovery from waste-see 2.3.5.6.

3.2.4 OTHER ENVIRONMENTAL INFORMATION DESCRIBING DIFFERENT WASTE CATEGORIES AND OUTPUT FLOWS

The parameters describing waste categories and other material flows are output flows derived from LCI. They are required and shall be included in the EPD as follows:

Other environmental information describing waste categories   Table 5

Parameter	Parameter unit expressed per functional/declared unit
Hazardous waste disposed	kg
Non-hazardous waste disposed	kg
Radioactive waste disposed	kg

Guidance:
Hazardous waste: hazardous waste that needs special treatment (excluding radioactive waste).
Non-hazardous waste: includes
Overburden: includes dust, spoil and other waste from raw materials extraction.
Municipal waste: waste treated in municipal disposal scheme
Radioactive waste: covers all level of radioactivity, essentially waste from nuclear power plants

Other environmental information describing output flows   Table 6

Parameter	Parameter unit expressed per functional/declared unit
Components for re-use	kg
Materials for recycling	kg
Materials for energy recovery	kg
Exported energy	MJ per energy carrier

Guidance: The parameters in Table 6 are calculated on the gross amounts leaving the system boundary when they have reached the end-of-waste state as described in 2.3.4.5.
These parameters may also have had impact allocated to them according to the allocation procedures defined in 2.3.5.3.
The declaration of "components for re-use" and "materials for recycling": fulfils the conditions of 4.2.5, end-of-life stage.
The parameter "Materials for energy recovery" does not include materials for waste incineration. Waste incineration is a method of waste processing and is allocated within the system boundaries. Waste incineration plants have a lower energy efficiency rate than power plants using secondary fuels. Materials for energy recovery are based on thermal energy efficiency rate of the a power plant not less than 60 % or 65 % for installations after 31st of December 2008 in order to be in line with the distinction made by the EC.
Exported energy relates to energy exported from waste incineration and landfill. Other energy exported from the system is not reported.
The characteristics that render waste hazardous are described in existing applicable legislation, e.g. in the European Waste Framework Directive 2008/98/EC.

3.3 SCENARIOS AND ADDITIONAL TECHNICAL INFORMATION

3.3.1 SUPPORTING PRODUCT LEVEL SCENARIOS

To enable building or construction works level assessment, scenarios can be used to model relevant life cycle stages. Information to support the calculation of scenarios that deal with any one or all of the life cycle stages of the construction product after manufacturing can be provided as part of the EPD, covering "transport to site, construction, use stage, end-of-life" (see Figure 3-2) Fel! Hittar inte referenskälla..

A scenario shall be realistic and representative of one of the most probable alternatives. (If there are, e.g. three different applications, the most representative one, or three scenarios can be declared). Scenarios shall not include processes or procedures that are not in current use or which have not been demonstrated to be practical.

Guidance: Energy recovery needs to be based on existing technology and current practice.
Guidance examples: *A recycling system is not practical if it includes a reference to a return system for which the logistics have not been established.*

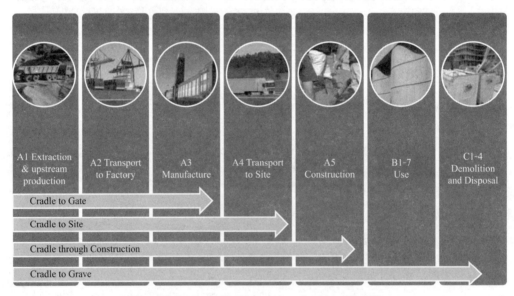

Figure 3-2. Life Cycle Stages described by this PCR.

3.3.2 GENERAL

Scenarios for further life cycle stages should support the application of product related data in the corresponding life cycle stage of the building assessment.

Additional technical information as defined in Table 7 to Table 12 supports the consistent development of scenarios by which the LCA derived parameters for further life cycle sta-

ges can be calculated and declared. To support the development of such scenarios, this PCR provides the information which can be provided optionally for products to enable consistent evaluation.

Additional technical information is declared in the life cycle section, to which it refers (e. g. technical information about the use of a product in the appropriate use stage modules.)

Any additional technical information shall be documented separately from the LCA derived parameters.

If additional technical information is not complete at the product level as specified in 3.3, this shall be stated.

Additional technical information in support of scenarios shall be realistic and representative of one of the most probable alternatives. (If there are, e. g. three different options, the most representative one, or all three options shall be declared). Additional technical information shall not include processes or procedures that are not in current use or which have not been demonstrated to be practical.

The following tables are not exhaustive with respect to examples or given units and parameters.

3.3.3 CONSTRUCTION PROCESS STAGE

3.3.3.1. A4, Transport to the construction site

If product stage A4 is declared, or if additional technical information is provided in the EPD for transport from the production gate to the construction site, the following information shall be provided to support development of the scenarios at the building or construction works level:

Typical transport to the construction site Table 7

Parameter	Parameter unit expressed per functional / declared unit
Fuel type and consumption of vehicle or vehicle type used for transport e. g. long distance truck, boat etc.	Litre of fuel type per distance or vehicle type,, Commission Directive 2007/37/EC (European Emission Standard)
Distance	km
Capacity utilisation (including empty returns)	%
Bulk density of transported products	kg/m^3
Volume capacity utilisation factor (factor: = 1 or <1 or ≥1 for compressed or nested packaged products)	Not applicable

Guidance: As an alternative to the bulk density the weight and volume of transported products may be specified.

With the bulk density and the volume capacity utilisation factor, (complex) logistic scenarios (e. g. taking onto account the type of vehicle, transport distance, empty returns) at the building level can be considered.

For the assessment at the building level more complex logistics may have to be considered.

3.3.3.2. A5, Installation in the building or construction works

If additional technical information is provided in the EPD for installation in or onto the building, the following information shall be provided to specify the product's installation scenarios or to support development of the scenarios describing the product's installation at the level of the building assessment:

This should cover, for example typical materials and energy required for formwork, pumping, installation, or treatment of concrete to assist curing (heating or cooling, use of water) and treatment to address the surface of the concrete. These are to be provided to allow the installation phase impacts to be calculated dependent on the actual building or construction works situation.

Typical Installation of the product in the building B1-B7 use stage　　　Table 8

Parameter	Unit (expressed per functional/declared unit)
Ancillary materials for installation (specified by material);	kg or other units as appropriate
Net consumption of fresh water for installation	m^3
Other resource use for installation	kg
Quantitative description of energy type (regional mix) and consumption during the installation process	kWh or MJ
Wastage of materials on the building site before waste processing, generated by the product's installation (specified by type)	kg
Output materials (specified by type) as result of waste processing at the building site e.g. of collection for recycling, for energy recovery, disposal (specified by route)	kg
Direct emissions to ambient air, soil and water	kg

3.3.3.3. B1-B5 use stage related to the building fabric

The provision of additional technical information to support the calculation of life cycle modules B1-B7 at the building level is optional and if included must be based on typical scenarios which must be described.

B1: Environmental aspects and impacts connected to the normal (i.e. anticipated) use of products, not including those related to energy and water use, which are dealt with in B6 and B7) e.g. release of substances from the facade, roof, floor covering, walls and other surfaces (interior or exterior) are reported as additional information (see 4.3).

B2-B5, if additional technical information is provided in the EPD for products requiring maintenance, repair, replacement, refurbishment the following information shall be provided to specify the scenarios or to support the development scenarios of these modules at the building level. Information given for Table 9 shall be consistent with the reference service life data given in Table 9:

use stage related to the building fabric   Table 9

Parameter	Parameter unit expressed per functional/declared unit
B2 Maintenance	
Maintenance process	Description or source where description can be found
Maintenance cycle	Number per RSL or year *
Ancillary materials for maintenance, (e. g. cleaning agent, specify materials)	kg /cycle,
Wastage material during maintenance (specify materials)	kg
Net fresh water consumption during maintenance	m^3
Energy input during maintenance (e. g. vacuum cleaning), energy carrier type e. g. electricity, and amount, if applicable and relevant	kWh
B3 Repair	
Repair process	Description or source where description can be found
Inspection process	Description or source where description can be found
Repair cycle	Number per RSL or year
Ancillary materials, (e. g. lubricant, specify materials)	kg or kg /cycle
Wastage material during repair, (specify materials)	kg
Net fresh water consumption during repair	m^3
energy input during repair (e. g. crane activity), energy carrier type e. g. electricity, and amount	kWh /RSL, kWh /cycle
B4 Replacement	
Replacement cycle	Number per RSL or year
energy input during replacement (e. g. crane activity), energy carrier type, (e. g. electricity) and amount if applicable and relevant,	kWh
exchange of worn parts during the product's life cycle, (e. g. zinc galvanised steel sheet), specify materials	kg
B5 Refurbishment	
refurbishment process	Description or source where description can be found
Refurbishment cycle	Number per RSL or year
energy input during refurbishment (e. g. crane activity), energy carrier type e. g. electricity, and amount if applicable and relevant,	kWh
material input forrefurbishment (e. g. bricks), including ancillary materials for the refurbishment process, (e. g. lubricant, specify materials)	kg or kg /cycle
Wastage material during refurbishment, (specify materials)	kg
Further assumptions for scenario development, (e. g. frequency and time period of use, number of occupants)	units as appropriate
*	not applicable if only B2 is declared

3.3.3.4. Reference service life

> Guidance: This PCR does not cover the use phase of the product and therefore the declaration of the RSL is not required. The EPD can however include information related to the calculation of the RSL as additional information.

The description of the reference service life may be based on data collected as average data or at the beginning or end of the service life. The reference conditions for achieving the declared technical and functional performance and the declared reference service life shall include the reference service life data as described in Table 10, where relevant:

Reference Service Life   Table 10

Parameter	Parameter unit expressed per functional/declared unit
Reference Service Life	Years
Declared product properties (at the gate) and finishes, etc.;	Units as appropriate
Design application parameters (if instructed by the manufacturer), including the references to the appropriate practices;	Units as appropriate
An assumed quality of work, when installed in accordance with the manufacturer's instructions;	Units as appropriate
Outdoor environment, (for outdoor applications), e.g. weathering, pollutants, UV and wind exposure, building orientation, shading, temperature;	Units as appropriate
Indoor environment (for indoor applications), e.g. temperature, moisture, chemical exposure;	Units as appropriate
Usage conditions, e.g. frequency of use, mechanical exposure;	Units as appropriate
Maintenance e.g. required frequency, type and quality and replacement of replaceable components.	Units as appropriate

3.3.4 END-OF-LIFE

If additional technical information is provided in the EPD about end-of-life processes, the following information shall be provided for all construction products to specify the end-of-life scenarios used or to support development of the end-of-life scenarios at the building level. Scenarios shall only model processes e.g. recycling systems that have been proven to be economically and technically viable.

Typical End-of-life   Table 11

Processes	Parameter unit expressed per functional/declared unit of components, products or materials (specified by type of material)
Collection process specified by type	kg collected separately
	kg collected with mixed construction waste

续表

Processes	Parameter unit expressed per functional/declared unit of components, products or materials (specified by type of material)
Recovery system specified by type	k g for re-use
	kg for recycling
	kg for energy recovery
Disposal specified by type	kg product or material for final deposition
Assumptions for scenario development, (e. g. transportation)	units as appropriate

3.4 AGGREGATION OF INFORMATION MODULES

Information modules A1, A2, and A3 may be aggregated or provided as 3 separate modules.

4 PROJECT VERIFICATION REPORT

4.1 GENERAL

The Project Verification Report is the systematic and comprehensive summary of the project documentation supporting the verification of an EPD. It is assembled by the party producing the EPD and submitted to a verifier nominated by the EPD Scheme. The Project Verification Report is used to demonstrate that the LCA based information and the additional information that are declared in the EPD meet the requirements of this PCR. It is essential in the Project Verification Report to demonstrate in a transparent way how the data and information declared in the EPD results from the LCA study.

The Project Verification Report has to be made available to the verifier with the requirements on confidentiality stated in EN ISO 14025.

The Project Verification Report is not communicated externally (other than to the Verifier) unless the manufacturer wants to do so.

> Guidance Examples: The Project Verification Report is produced at the same time and by the same party as the EPD and provided by them to the verifier to facilitate verification. The format of the Project Verification Report may be determined by the program operator.

4.2 LCA-RELATED ELEMENTS OF THE PROJECT VERIFICATION REPORT

The results, data, methods, assumptions and limitations and conclusions of the LCA shall be completely and accurately reported without bias. They shall be transparent and presented in sufficient detail to allow independent verification and to permit an understanding of the complexities and trade-offs inherent in the LCA. The report should also allow the results and interpretation to be used in support of the data and additional information made available in the respective EPD.

The Project Verification Report shall give the following:

a) *General aspects:*
   1) *commissioner of the LCA study, internal or external practitioner of the LCA study;*
   2) *date of report;*
   3) *statement that the study has been conducted according to the requirements of this standard;*

b) *Goal of the study:*
   1) *reasons for carrying out the study and its intended application and audience,*

i. e. providing information and data for an EPD for business-to-business and /or business-to-consumer communication;

c) *Scope of the study*:

1) *declared unit, including*:
    i) *definition, including relevant technical specification (s) (see 2.3.2)*,
    ii) *calculation rule for averaging data e. g. when the declared /functional unit is defined for*:
        1. *the same product produced at different production sites (see 2.5.2) or*
        2. *a group of similar products produced by different suppliers (see 2.5.2)*.
2) *system boundary (see 2.3.4) according to the modular approach, including*:
    i) *omissions of life cycle stages, processes or data needs*;
    ii) *quantification of energy and material inputs and outputs, taking into account how plant-level data is allocated to the declared products*; *and*
    iii) *assumptions about electricity production and other relevant background data*;
3) *cut-off criteria for initial inclusion of inputs and outputs (see 2.3.5.3), including*:
    i) *description of the application of cut-off criteria and assumptions*;
    ii) *list of excluded processes*;

d) *Life cycle inventory analysis*:

1) *qualitative /quantitative description of unit processes necessary to model the life cycle stages of the declared unit, taking into account the provisions of EN ISO 14025 regarding data confidentiality*;

> Guidance: ISO 14025 states the following:
> Product-specific data are often confidential because of
> - competitive business requirements,
> - proprietary information covered by intellectual property rights, or
> - similar legal restrictions.
>
> Such confidential data are not required to be made public. The declaration typically only provides data aggregated over all or relevant stages of the life cycle. Business data identified as confidential that is provided for the independent verification process shall be kept confidential.

2) *sources of generic data or literature used to conduct the LCA*;
3) *validation of data (see 2.4.3,* **Fel! Hittar inte referenskälla.** *and 2.4.4), including*:
    i) *data quality assessment*; *and*
4) *treatment of missing data*;
5) *allocation principles and procedures (see* **Fel! Hittar inte referenskälla.** *), including*:
    i) *documentation and justification of allocation procedures*; *and*
    ii) *demonstration that uniform application of allocation procedures*;

e) Life cycle impact assessment (see 3.2 Fel! Hittar inte referenskälla.):
   1) the LCIA procedures, calculations and results of the study;
   2) the relationship of the LCIA results to the LCI results;
   3) reference to all characterization models, characterization factors and methods used, as defined in this European Standard;
   4) a statement that the LCIA results are relative expressions and do not predict impacts on category endpoints, the exceeding of thresholds, safety margins or risks;

f) Life cycle interpretation:
   1) the results;
   2) assumptions and limitations associated with the interpretation of results as declared in the EPD, both methodology and data related;
   3) the variance from the mean of each LCIA results should be described, if generic data are declared from several sources or for a range of similar products;
   4) data quality assessment;
   5) full transparency in terms of value-choices, rationales and expert judgments.

4.3 DOCUMENTATION ON ADDITIONAL INFORMATION

The Project Verification Report shall include any documentation to support any additional environmental information declared in the EPD as required in this standard.

> Guidance: Such documentation could include the following, e.g. as copies or as links to references:
> - laboratory results /measurements for the content declaration;
> - laboratory results /measurement of functional /technical performance;
>
> documentation on declared technical information on life cycle stages that have not been considered in the LCA of the construction product and that will be used for the assessment of buildings (e.g. transport distances, RSL, energy consumption during use, cleaning cycles, etc.).

4.4 DATA AVAILABILITY FOR VERIFICATION

To facilitate verification it is considered good practice to make the following information available to the verifier, taking into account data confidentiality according to ISO 21930: 2007, 7.4 and 9.1:

- analysis of material and energy flows to justify their inclusion or exclusion;
- quantitative description of unit processes that are defined to model processes and life cycle stages of the declared unit;
- attribution of process and life cycle data to datasets of an LCA-software (if used);

- LCIA results per modules of unit processes, e.g. structured according to life cycle stages;
- LCIA results per production plant/product if generic data is declared from several plants or for a range of similar products;
- documentation that substantiates the percentages or figures used for the calculations in the end-of-life scenario;
- documentation that substantiates the percentages and figures (number of cycles, prices, etc.) used for the calculations in the allocation procedure, if it differs from the PCR.

4.5 VERIFICATION AND VALIDITY OF AN EPD

After verification an EPD is valid for a 5 year period from the date of issue.

After 5 years, the EPD shall be reviewed. An EPD shall only be reassessed and updated as necessary to reflect changes in technology or other circumstances that could alter the content and accuracy of the declaration. An EPD does not have to be reassessed after 5 years if the underlying data has not changed significantly. In either case, the EPD will need to be verified again to have a further 5 year period of validity.

The process for verification and establishing the validity of an EPD shall be in accordance with EN ISO 14025 and ISO 21930.

Guidance: A reasonable change in the environmental performance of a product to be reported to the verifier is +/- 10% on any one of the declared parameters of the EPD (see Clause 7). Such a change may require an update of the EPD.

5 TERMS AND DEFINITIONS

The following definitions are provided for terms used in the PCR and Guidance.

1.1. *additional technical information*: *information that forms part of the EPD by providing a basis for the development of scenarios*

> Guidance example: An example of the type of additional technical information which could be provided in an EPD for concrete is the diesel required to pump 1m^3 of concrete 10 metres vertically upward. This information could then be used by someone developing a Building LCA to generate the impact, by taking the volume of concrete and the average height that it needs to be raised to calculate the total amount of diesel needed on site for pumping concrete.

1.2. alternative fuels: alternative fuels (AF) are typically derived from wastes and therefore, without this use, would have to be disposed of in some other way, usually by landfilling or incineration. Alternative fuels serve as a substitute for conventional fossil fuels. They include fossil fuel-based or non biogenic fractions, such as, waste oil and plastics, and biomass fractions, such as waste wood and sewage sludge.

1.3. *ancillary material*: *input material or product that is used by the unit process producing the product, but which does not constitute part of the product* [EN ISO 14040: 2006]

> Guidance example: An example of an ancillary material used in the production of concrete would be formwork, as it does not constitute part of the final product.

1.4. *average data*: *data representative of a product, product group or construction service, provided by more than one supplier*

> Guidance: The product group or construction service can contain similar products or construction services. For example, an EPD could be provided for individual products such as a particular compressive strength of hollow or solid dense block products, or for dense concrete blocks as a product group, covering all the relevant products within that group.
> Note that EPD may be produced for one or more products produced by one supplier (see 2.5.1), or for a product, products or product group produced by more than one supplier (eg. a trade association) (see 2.5.2), and that different rules apply to these situations and how the results should be reported.

1.5. **building**: construction works that has the provision of shelter for its occupants or contents as one of its main purposes and is usually enclosed and designed to stand permanently in one place [ISO 6707-1: 2004]

1.6. **by-product**: secondary substances produced by an industrial process (ISO/DIS 13315-1)

> Guidance example: An example of a by-product would be fuel ash produced within a coal fired power station. By implication, a by-product is a waste which has been recycled or a co-product with very low value. See also 2.9 Co-product.

1.7. *comparative assertion*: *environmental claim regarding the superiority or equivalence of one product versus a competing product that performs the same function* [EN ISO 14044: 2006]

> Guidance there are rules within the Life Cycle Assessment Standards (ISO 14040: 2006, 14044: 2006) which prevent the use of comparative assertions unless the underlying life cycle assessment study has been peer reviewed by experts. This ensures that the goal and scope of the study supports the comparison of the products undertaken and that there is robust and consistent data used to support the comparison, and the products and/or services compared have the same functionality. EPD in themselves are not comparative assertions, as they only provide information on the impact of the product covered by the EPD. EPD can only be used to compare solutions when the function of the products within the building or construction works, and their impact over the life cycle has been considered. This is covered further in Section 4.3.

1.8. **concrete demolition waste**: concrete generated in demolition of concrete structures (ISO/DIS 13315-1)

1.9. **construction element**: part of a construction containing a defined combination of products

> Guidance example: An example of a construction element would be a concrete ground floor slab, or a concrete blockwork cavity wall.

1.10. *construction product*: *item manufactured or processed for incorporation in construction works*

> Guidance: construction products are supplied by a single responsible body-where individual construction products are supplied separately to site and combined together there, they are known as a construction assembly, assembled system or construction element.

1.11. *construction service*: *activity that supports the construction process or subsequent maintenance*

> Guidance example: An example of a construction service would be the excavation of foundations or erection of formwork or scaffolding.

1.12. **construction works**: everything that is constructed or results from construction operations

> Guidance: This covers both building (3.3) and civil engineering works, and both structural and non-structural elements. [Adapted from the definition in ISO 6707-1: 2004].

1.13. *co-product*: *any of two or more marketable materials, products or fuels from the same unit process, but which is not the object of the assessment*

> Guidance: Co-product, by-product and product have the same status and are used for identification of several distinguished flows of products from the same unit process. From co-product, by-product and product, waste is the only output to be distinguished as a non-product.

1.14. *declared unit*: *quantity of a construction product for use as a reference unit in an EPD for an environmental declaration based on one or more information modules*

> Guidance examples: Mass (kg) or volume (m^3). [Adapted from ISO 21930: 2007] Other examples of declared units used of EPD of concrete are provided in 2.3.2

Fel! Hittarintereferenskälla.

1.15. *direct land use change (dLUC)*: *change in human use or management of land at the location of the production, use or disposal of raw materials, intermediate and final products or wastes in the product system (From ISO 14067 CD3)*

> Guidance Examples: Opening or closing quarries or factories may result in land use change, eg. from forest to quarry, with consequential impacts in terms of greenhouse gas emissions. Section 2.6.2 covers how any significant land use change should be considered.

1.16. *drainage basin*: *area that captures water in any form, such as rain, snow or dew and drains it to a common water body (from WD ISO 14046)*

> Guidance: Drainage basins can be identified by tracing a line along the highest elevations between two areas on a map, i.e. drainage divides, that determine the direction the water will flow. Sometimes the terms "watershed", "drainage area", "catchment basin", "catchment area" or "river basin" are used for this concept. Groundwater drainage basin not necessarily reflects surface drainage basin.

1.17. *end-of-waste state*: *Point at which waste can no longer be considered waste.*

> Guidance: A definition of the end-of-waste state is provided in 2.3.4.5. This is drawn from the European Waste Framework Directive. However, the end-of-waste state as defined does not ensure a consistency of application which WBCSD CSI would like to see, to ensure comparability and consistency of EPD. This is demonstrated because, a) there are differences in interpretation and implementation of this directive in each European state, b) the directive does not pertain outside of Europe, c) there are differences in the economic values of wastes, by-products and co-products in different locations, and d) different states have introduced protocols to define the qualities of material which has reached the end-of-waste state.

For the purposes of this PCR, WBCSD CSI are therefore providing guidance on the point at which wastes should be considered to reach the end-of-waste state wherever they are produced, and whatever their legal status in that location. This guidance is based on a "conservative" principle-so that the end-of-waste state is identified, if there is more than one point, at the point where the benefit to concrete producers is minimised, rather than maximized.

Guidance examples:

An example of the end-of-waste state for concrete demolition waste is the point in the process from the demolition of the building through to the eventual use of the recycled concrete aggregate, at which the material moves from being a waste to a product. In this instance, it is defined as the point at which the material has been crushed and can be used as a product.

An example of the end-of-waste state of waste tyres, used as a fuel, is as follows: waste tyres, put whole into the cement kiln-product is a waste when it enters the kiln. waste tyres, shredded and steel reinforcement recovered before putting into the cement kiln-product reaches end-of-waste state when it has been prepared as a secondary fuel.

1.18. **environmental declaration**: claim which indicates the environmental aspects of a product or service ISO 14020: 2001 sets out 3 types of environmental declarations:

-Type Ⅰ, known as ecolabels, prescribed in ISO 14024: 2006, and covering schemes such as the EU Ecolabel or Blauer Engel.

-Type Ⅱ, self declared environmental claims, described in ISO 14021: 2006, and covering claims such as recycled content, recyclable, reduced energy consumption etc.

-Type Ⅲ, Environmental Declarations, described in ISO 14025: 2006 and covering construction EPD as further described in ISO 21930: 2007.

1.19. **environmental performance**: *performance related to environmental impacts and environmental aspects* [*ISO 15392: 2008*]. [*ISO 21931-1: 2010*]

Guidance example: The environmental performance of the building is the environmental impact that it has as a result of its construction, operation, demolition and disposal. It is important to understand that different materials can have considerably different effects on the environmental performance of a building. For example, in terms of building operation, concrete materials can impact on the thermal mass or thermal inertia of the building, the air-tightness of the building and the thermal resistance of the building. Additionally, construction products such as concrete can affect the design of other parts of the building-for example the weight of materials used in a wall could affect the mass of foundations required; the thickness of the wall could affect the building footprint or the useable floor area of the building; a material which has a fair-faced finish does not have to use an additional surface finish treatment such as plasterboard and paint.

> Impacts associated with the use of construction materials and services within the building or construction works are calculated on the basis of information provided in EPD. Most countries have also defined standardized mechanisms to calculate or estimate the impact of building operation from energy and water use
> -these may however not equate to actual energy and water consumption due to differences in the real and theoretical operation of the building.

1.20. **Environmental Product Declaration (EPD)**: a Type III environmental declaration
1.21. *fossil water*: *water that has remained sealed in an aquifer for a long period of time (hundreds to millions years of time) (from WD ISO 14046). Also known as paleowater.*
Fossil water is a non-renewable resource or renewable only over long geological periods. Water can rest underground in " fossil aquifers" for thousands or even millions of years. When changes in the surrounding geology seal the aquifer off from further replenishing from precipitation, the water becomes trapped within, and is known as fossil water.
1.22. *functional equivalent*: *quantified functional requirements and/or technical requirements for a building or an assembled system (part of works) for use as a basis for comparison*

> Guidance: as this PCR only covers the cradle to gate or site stages of the construction life cycle, it is not possible to consider a functional equivalent, as this needs to take into account the full life cycle including the use stage and end of life.

1.23. *functional unit*: *quantified performance of a product system for use as a reference unit* [EN ISO 14040: 2006]

> Guidance: as this PCR only covers the cradle to gate or site stages of the construction life cycle, it is not possible to define a functional unit, as this needs to take into account the full life cycle including the use stage and end of life. As a cradle to gate or cradle to site EPD, the declared unit is all that can be used.

1.24. **greenhouse gas removal, GHG removal**: total mass of a greenhouse gas removed from the atmosphere over a specified period of time [ISO 14064-1: 2006, 2.6]
1.25. **groundwater**: water contained in aquifers which are strata in which the water is contained in the porous voids (from WD ISO 14046)
1.26. **independent verifier**: Independent verifiers, whether internal or external to the organization, shall not have been involved in the execution of the LCA or the development of the declaration, and shall not have conflicts of interests resulting from their position in the organization. (ISO 14025 2006 8.2.1)

> Guidance: EPD require independent verification. The verifier could be from the manufacturer, an LCA practitioner that has undertaken the EPD for the manufacturer, could be appointed by an EPD program or chosen from a list of approved verifiers, but it is essential that the verifier can demonstrate their independence from the study, or any conflict of interest in relation to their position vis a vis those undertaking the study-for example that they do not manage the staff undertaking the study or have responsibility for the quality of their work. From this point of view, an independent verifier appointed by an EPD program or chosen from a list of approved verifiers offers the most confidence in terms of external communications.

1.27. *information module*: *compilation of data to be used as a basis for a Type III environmental declaration covering a unit process or a combination of unit processes that are part of the life cycle of a product* [EN ISO 14025: 2010]

> Guidance example: Data on the typical impacts of transport to site for a concrete product may be provided as an information module (A4 within the terminology of EN 15804). This provides the LCA data (environmental impact and inventory indicators) and the data underlying the scenario used, for example the type of vehicle, distance travelled, load carried, empty returns etc.

1.28. *life cycle assessment (LCA)*: *compilation and evaluation of the inputs, outputs and the potential environmental impacts of a product system throughout its life cycle* [EN ISO 14044: 2006]

> Guidance: LCA is governed by the ISO standards ISO 14040: 2006 and ISO 14044: 2006 which set out the framework for undertaking any LCA studies.

1.29. *life cycle inventory analysis (LCI)*: *phase of life cycle assessment involving the compilation and quantification of inputs and outputs for a product throughout its life cycle* [EN ISO 14040: 2006]

> Guidance: Life Cycle Inventory is a list of all the raw material resources, at the point at which they are extracted from nature, i.e. crude oil, iron ore, limestone rock etc, and all the emissions to air, water or land, associated with a given system. Each item in the inventory is known as a "burden". The Inventory is produced by tracing all resource use and emissions "upstream" through the supply chain from manufacture to where man first has an influence on the environment, and "downstream", eg. through waste treatment and disposal. Life Cycle Inventory data needs further processing, known as Impact Assessment, to "classify" the environmental impacts each resource use or emission cause, and to "characterize" the size of the impact from each burden.

混凝土 EPD 评价采用的 PCR 示例

1.30. *non-renewable energy*: energy from sources which are not defined as renewable energy sources

> Guidance examples: fossil fuels and uranium are examples of non-renewable energy sources which cannot be replenished on a human timescale.

1.31. *non-renewable resource*: resource that exists in a finite amount that cannot be replenished on a human time scale [ISO 21930: 2007]

> Guidance examples: limestone, sand and gravel, marine dredged aggregate and other extracted stones are examples of non-renewable resources.

1.32. *performance*: expression relating to the magnitude of a particular aspect of the object of consideration relative to specified requirements, objectives or targets

1.33. *product category*: group of construction products that can fulfil equivalent functions

> Guidance: This PCR covers the product category unreinforced concrete, and includes various products such ready mix concrete, concrete blocks, mortars and renders.

1.34. *product category rules* (*PCR*): set of specific rules, requirements and guidelines for developing Type III environmental declarations for one or more product categories

> Guidance: A PCR is provided to ensure that manufacturers using the same PCR can produce Environmental Product Declarations (EPD) which are robust, independently verified, consistent and can be used at the building or construction works level to assess the impact of the building or construction works to compare different products taking account of their influence on the building or construction works.
> This PCR for unreinforced concrete, by providing detailed rules, requires and guidelines, and by giving explanations and examples, should minimize the opportunity for interpretation within the document, and is intended to ensure that anyone using it and assessing the same product would produce a repeatable and comparable result.

1.35. *product system*: collection of unit processes with elementary and product flows, performing one or more defined functions, and which models the life cycle of a product [EN ISO 14040: 2006]

> Guidance: The cradle to gate product system for ready-mix concrete for example would take into account the manufacture of the raw materials required, transport to the ready mix plant and production of the concrete. To extend the product system to the site, transport processes and associated washing of lorries must be considered for example. To further include installation on site, any processes associated with pumping the concrete, use of formwork, any activities to ensure curing or to produce the required surface qualities must be included in the product system. This PCR does not extend beyond installation on site, but to model the full life cycle, maintenance and refurbishment processes, and demolition and any associated waste treatment and disposal processes would need to be considered.

1.36. *programme operator*: body or bodies that conduct a Type Ⅲ environmental declaration programme

> Guidance: A programme operator can be a company or a group of companies, industrial sector or trade association, public authorities or agencies, or an independent scientific body or other organization. Guidance Examples: Programme Operators producing EPD for construction products include the Insitut Bau und Umwelt (IBU) (www.bau-umwelt.de), the International EPD® Scheme (www.environdec.com) and certification bodies such as Underwriters Laboratories (http://www.ulenvironment.com) and BRE Global (www.greenbooklive.com).

1.37. *reference service life* (RSL): service life of a construction product which is known to be expected under a particular set, i.e., a reference set, of in-use conditions and which may form the basis of estimating the service life under other in-use conditions [ISO 21930: 2007]

1.38. *reference service life data* (RSL data): information that includes the reference service life and any qualitative or quantitative data describing the validity of the reference service life

> Guidance example: Typical data describing the validity of the RSL include the description of the component for which it applies, the reference in-use conditions under which it applies, and its quality. [ISO 15686-8: 2008. Guidance: as the PCR only applies for cradle to gate or cradle to site assessments, it is not necessary within the EPD to consider or report the reference service life.

1.39. *renewable energy*: energy from renewable non-fossil sources

> Guidance: energy which is generated from renewable resources is considered to be renewable.
> Guidance examples: Wind, solar, aerothermal, geothermal, hydrothermal and ocean energy, hydropower, biomass, landfill gas, sewage treatment plant gas and biogases.

1.40. *renewable resource*: resource that is grown, naturally replenished or naturally cleansed, on a human time scale

> Guidance: A renewable resource is capable of being exhausted, but may last indefinitely with proper stewardship. Examples include: trees in forests, grasses in grassland, fertile soil. [ISO 21930: 2007] If a resource is harvest unsustainably then it cannot be regarded as a renewable resource-for example timber from illegal logging is not a renewable resource.

1.41. **returned water**: water that returns after use to a water body (from WD ISO 14046)

> Guidance: The quality, quantity, temperature and point of return to a water body compared to pre-withdrawal conditions can be different.
> Guidance Example: Water which is withdrawn from a river or lake for use in a cooling system may be returned after use back to the same river or lake. This is returned water.

1.42. *scenario*: *collection of assumptions and information concerning an expected sequence of possible future events*

> Guidance: Scenarios are required to enable assessment of stages of the life cycle which will occur in the future. This is because the future is unknown. Within EPD, scenarios should be based on what is known to currently occur now, rather than on what is estimated will happen in the future. For example, a scenario for the life cycle stage A4-transport to site, might provide the typical distance that the product travels to the customer in a given country, the mode of the transport (road/rail etc), the vehicle type, load and empty lading %. The scenario should explain the basis of this data-for example that it is the mean or modal average data. This scenario information can be used to calculate the impacts of this scenario for inclusion within the EPD, or can just be included as additional information. For those using the EPD, the scenario can be reviewed to see if it is appropriate for the actual situation being considered. If the site is double the distance for example, then the impacts, if calculated, can be doubled. Alternatively, if the impact of transport is seen to be insignificant compared to the impacts of manufacture, then it could be ignored.

1.43. *secondary fuel*: *fuel recovered from previous use or from waste which substitutes primary fuels*

> Guidance: *Processes providing a secondary fuel are considered from the point where the secondary fuel enters the system from the previous system.*
> *Any combustible material recovered from previous use or from waste from the previous product system and used as a fuel in a following system is a secondary fuel.*
> *Guidance examples for primary fuels are*: *coal, natural gas, biomass, etc.*
> *Guidance examples for secondary fuels recovered from previous use or as waste are*: *used solvents, used wood, used tyres, used oil, waste animal fats.*
> It should be noted that secondary fuels may not have reached the end-of-waste state before use-for example waste tyres recovered for use whole within a cement kiln. The term, secondary fuel, therefore does not have any impact on the treatment of the fuel. Use of secondary fuels (whether they have reached the end-of-waste state or not, are reported as an inventory indicator.

1.44. *secondary material*: material recovered from previous use or from waste which substitutes primary materials

> *Guidance*: Secondary material is measured at the point where the secondary material enters the system from another system.
> Materials recovered from previous use or from waste from one product system and used as an input in another product system are secondary materials-at the point of use, they may not have reached the end-of-waste state.
> Guidance Examples: Examples for secondary materials (to be measured at the system boundary) are crushed concrete, glass cullet, pulverized fuel ash.

1.45. *specific data*: data representative of a product, product group or construction service, provided by one supplier

> Guidance Examples: the data provided by a factory regarding their actual energy consumption, raw materials inputs, transport of raw materials, amount and destination of wastes, emissions and outputs are specific data. This is sometimes also known as primary data. The other type of data is secondary or generic data-this is data provided by a trade association, from a national survey or report or industry report, or a database and is based on data from more than one supplier or from an estimation of the data.

1.46. surface water: all water you can see as overland flow, rivers and lakes excluding sea water (from WD ISO 14046)

1.47. *third party*: person or body that is recognized as being independent of the parties involved, as concerns the issues in question

> *Guidance*: "Parties involved" are usually supplier ("first party") and purchaser ("second party") interests. [*EN ISO 14024: 2000*]
> Other ISO standards, for example, ISO 9001, talk about "second party" being any party with an interest in the supplier, for example, a stakeholder. Third party verification is only required for business to consumer communication-for business to business communication (i.e. between a supplier and specifier, or supplier and contractor or supplier and client), the only requirement is that the verifier is competent and independent.

1.48. *type III environmental declaration*: environmental declaration providing quantified environmental data using predetermined parameters and, where relevant, additional environmental information, which has been independently verified.

> *Guidance*: The calculation of predetermined parameters is based on prEN 15804, which itself has been based on the ISO 14040 series of standards, which is made up of ISO 14040, and ISO 14044, and ISO 21930.

1.49. **upstream, downstream process**: process (s) that either precedes (upstream) or follows (downstream) a given life cycle stage.

> Guidance: A manufacturer provides specific data about their use of energy, input materials and wastes.
> Upstream data is used to describe all the processes occurring before the manufacturer's process, for example the extraction of resources, the generation of energy and manufacture of input materials and their transport.
> Downstream data is used to describe all the processes occurring after the manufacturer's process, for example transport, treatment and/or disposal of wastes, transport of products, and their use and end of life product stages.

1.50. **waste**: substance or object which the holder discards or intends or is required to discard

> Guidance: this definition is adapted from the definition in the European Waste Framework Directive 2008/98/EC.
> Guidance Examples: to be covered once agreed end-of-waste state.

1.51. **water consumption**: water withdrawal where release back to the source of origin does not occur, e.g. because of evaporation, evapotranspiration, product integration or discharge into a different drainage basin or the sea (from WD ISO 14046)

> Guidance: The granularity of the source of origin needs to be defined in the goal and scope. Guidance Examples:
> Evaporation: Water is withdrawn and subsequently evaporates into the atmosphere, for example when it is lost from a cooling operation; when it is used for dust suppression of stockpiles or quarries. Once the water has evaporated, it becomes part of the wider water cycle and could return to a waterbody anywhere in the world.
> From this point of view, it is considered as water consumed.
> Evapotranspiration: Water is taken up by plants or other biomass and is released to the atmosphere via evapotranspiration. Natural evapotranspiration, for example from wild planting as a result of rainfall, is not considered water consumption. However the use of water for irrigation and the water lost from evapotranspiration from planting (eg. lawns) are considered water consumption. This type of water consumption is unlikely to be an issue for the concrete industry.
> Product integration (sometimes known as embedded water): this is the water which is remains in the product when it leaves the factory gate. For concrete, this is both the water making up the moisture content and the water which is chemically integrated with the cement. To all intents and purposes, this water is no longer available and is therefore considered as consumed.

1.52. **water scarcity**: parameter which describes to which extent water use exceeds the natural generation of water in an area, e.g. a drainage basin (from WD ISO 14046)

> Guidance: Water scarcity occurs where there are insufficient water resources to satisfy environmental water requirements and human water demands.
> Guidance Examples http://www.fao.org/nr/water/art/2007/scarcity.html shows an illustration of water scarcityacross the globe.

1.53. **water use**: any use of water by human activity, including agriculture, industry, energy production, public sector and households, including in-stream or in-situ uses such as fishing, recreation, transportation and waste disposal (from WD ISO 14046)

> Guidance: Water use encompasses any process or activity which makes use of water. This can cover "off-stream" water use, where the water is removed from its watercourse, for example water which is used by an activity, such as the water which is used to mix concrete, or to clean truck wheels, or to dampen materials such as sand heaps or the use of water in a cooling or cleaning process, where the water is extracted and recirculated and may be returned to source. It also covers "in-stream water", which is water used within a watercourse, for example the use of water as a transport media for boats; the use of water by a fisherman; the use of water in a process such as hydroelectric power generation, both in providing potential energy and by turning the turbines; or the use of water to dilute effluents, for example in a river.
> Note that water use and water consumption are very different.

1.54. **water withdrawal**: anthropogenic removal of water from any water body, either permanently or temporarily (from WD ISO 14046)

> Guidance: Sometimes, the term "water abstraction" is used for this concept. Water withdrawal encompasses consumption and borrowing (where water is withdrawn but returned to the same water body with no change in quality) and where withdrawal water is returned to the same water body with a change in quality.
> Water withdrawal is equivalent to "offstream" water use.

6 ABBREVIATIONS

EPD	Environmental product declaration
GHG	Greenhouse gas
ISO	International Standards Organisation
PCR	Product category rules
LCA	Life cycle assessment
LCI	Life cycle inventory analysis
LCIA	Life cycle impact assessment
RSL	Reference service life

GreenFormat 第一版和第二版（草案）示例

GreenFormat

Structured Format for Information to Support Sustainable Design and Product Choices

Copyright © 2015. U. S. copyright held by CSI, Alexandria, VA.

All rights reserved, including World Rights and Electronic Rights. Except as permitted under the United States Copyright Act of 1976 and all subsequent amendments, no part of this publication may be reproduced or distributed in any form or by any means, or stored in a database or retrieval system, without the prior written permission of the publisher, The Construction Specifications Institute (CSI).

For use of any portion of GreenFormat in commercial applications, educational programs, or publications, please contact CSI at csi@csinet.org to obtain copyright license.

GreenFormat: Structured Format for Information to Support Sustainable Design and Product Choices
ISBN 978-0-9845357-8-1 $ 00.00Retail

Cataloging-in-Publication Data is on file with the Library of Congress

CSI gives acknowledgement and thanks to the members of the GreenFormat Task Team for their work:

George Middleton, CSI, AIA, LEED AP BD+C, GGA, Chair
Paul Sternberg, CSI, CCS, CCCA, LEED AP, NCARB
Chris Hsieh, LEED AP
Chris Dixon, CSI, CCS, LEED AP
Susan Kaplan, CSI, CDT, LEED AP
Renee Doktorczyk, CSI, CCS, FAIA, SCIP
Wade Bevier, FCSI, CCS, LEED AP BD+C, GGP
Andy McIntyre, CSI, CCPR, CIEA, LEED GA
Dawn Spears, CSI, CCPR, LEED AP BD+C, GGP, EDAC
Steven Thorsell, CSI, AIA
Mike MacVittie, Technical Committee Liaison

Introduction

CSI's GreenFormat™ is a standardized structure for organizing sustainable information elements associated with materials, products, systems, and technologies used in the built environment. By using this standardized format, manufacturers are assisted in identifying key product characteristics and providing designers, constructors, and building operators with information needed to help meet sustainable design and operation goals. The identification of the criteria, standards, and applicable certifications using GreenFormat provides designers, constructors, and building operators an effective way to evaluate the sustainable characteristics of materials, products, and processes.

The *GreenFormat* Task Team has organized the structure into specific categories in an effort to make the process of understanding and using sustainability related product information effective. Information classified according to the GreenFormat system is grouped into five categories, each containing sub-categories and classifications.

The first three categories allow suppliers and manufacturers to identify information specific to their products. These categories include the following:

1.0   Product General Information
2.0   Product Properties
3.0   Product Life Cycle

The last two categories focus on information about the manufacturer, which may include environmental policies, environmental or life cycle assessment initiatives, and social responsibility initiatives. These categories are as follows:

4.0   Manufacturer Sustainability Policies
5.0   Manufacturer Support Documentation

GreenFormat's flexible structure is designed to be adaptable to anticipated changes in the industry. As sustainability issues and product selection criteria evolve, new topics may be added in the appropriate category, and existing topics that become obsolete, redefined, or changed can be revised or removed as necessary. The structure is designed to support specification writers and the project construction documents they produce by providing the information needed for the four different methods of specifying: reference standard, performance, descriptive, and proprietary.

One way to understand the structure and purpose of GreenFormat and how it supports the specification development process is to consider the fact that specifications are typically written with a project requirement stated as a salient feature followed by a value for that salient feature. For example, a salient feature such as "VOC content" included in a project specification would be accompanied by a value such as "5 grams/liter." This would be one element of information provided by a product manufacturer under a category heading defined under GreenFormat.

The organization of GreenFormat into specific categories or classifications of information is not intended to imply a hierarchy or relative level of importance of the information significance within a category. The user of the information classified according to GreenFormat maintains the responsibility to evaluate and determine the suitability of the information provided. The user of *GreenFormat* determines project specific requirements and then prioritizes the sustainability criteria by which products will be evaluated. Many evaluation and rating systems exist to assist owners, designers, and specifiers in determining the relative contributions of different products to sustainability. Product data developers can organize their data reporting according to GreenFormat categories while retaining their own proprietary features.

One of the principal goals of GreenFormat is for product-related information to be transparent and verifiable. GreenFormat also emphasizes the importance of objective, science-based, and widely recognized standards and evaluation criteria arrived at through consensus standard-setting processes. GreenFormat is also designed to work in conjunction with MasterFormat™ and Section/PageFormat™ and can be applied to all construction products and categories. (See the CSI publications by the same titles for further information.)

GreenFormat 第一版和第二版（草案）示例

GreenFormat Outline Structure

1.0 PRODUCT GENERAL INFORMATION

    1.1 PRODUCT INFORMATION

        1.1.1 MasterFormat Number

        1.1.2 OmniClass Classification

        1.1.3 UniFormat Classification

        1.1.4 Product Identification

        1.1.5 Product Description

        1.1.6 Product Photo(s) or Drawing(s)

    1.2 MANUFACTURER INFORMATION

        1.2.1 Company Name

        1.2.2 Subsidiary Name(s)

        1.2.3 Address

        1.2.4 Contact Information

        1.2.5 Manufacturer's Authorized Representative

        1.2.6 Technical Sales and Product Representatives

2.0 PRODUCT PROPERTIES

    2.1 Sustainable Standards and Certifications

        2.1.1 Third-Party Certification

            2.1.1.1 Whole Product Sustainability

            2.1.1.2 Single Attribute

            2.1.1.3 Other Certification Categories

        2.1.2 Second Party Certification

            2.1.2.1 Whole Product Sustainability

            2.1.2.2 Single Attribute

            2.1.2.3 Other Certification Categories

        2.1.3 Self-Declaration of Compliance

            2.1.3.1 Whole Product Sustainability

            2.1.3.2 Single Attribute

            2.1.3.3 Other Certification Categories

    2.2 SUSTAINABLE FEATURES

- 2.2.1 Rating System Credits
- 2.2.2 Recycled Content
- 2.2.3 Rapidly Renewable Materials
- 2.3.4 Reused Materials
- 2.2.5 Emissions
- 2.2.6 Chemical Composition

3.0 PRODUCT LIFE CYCLE

3.1 LIFE CYCLE ASSESSMENT (LCA)
- 3.1.1 Goal and Scope of LCA
- 3.1.2 Life Cycle Inventory (LCI) Analysis Phase
- 3.1.3 Life Cycle Impact Assessment (LCIA) Phase
- 3.1.4 Limitations of the LCA

3.2 LIFE CYCLE INPUTS & OUTPUTS
- 3.2.1 Energy Inputs
- 3.2.2 Material Inputs
- 3.2.3 Air Emissions
- 3.2.4 Water Emissions
- 3.2.5 Waste Disposal

3.3 INDUSTRIAL PRODUCTION SYSTEM
- 3.3.1 Raw Material Extraction
- 3.3.2 Manufacturing and Production
- 3.3.3 Transportation and Distribution
- 3.3.4 Construction, Operations, and Maintenance
- 3.3.5 Recycling and Waste Management

4.0 MANUFACTURER SUSTAINABILITY POLICIES

4.1 ENVIRONMENTAL STEWARDSHIP
- 4.1.1 Description of Environmental Policies or Programs
- 4.1.2 Facilities Certified under Green Building Rating Systems
- 4.1.3 Awards or Recognition for Environmental Stewardship
- 4.1.4 Participation in Voluntary Environmental Impact Reduction Programs
- 4.1.5 Design for Environment Strategies
- 4.1.6 Standards and Certifications

4.2 CORPORATE GOVERNANCE
- 4.2.1 Employment Policies

4.2.2 Employer Responsibilities
4.2.3 Community Engagement at Plant Level
4.2.4 Financial Leadership at Corporate Level

4.3 MANUFACTURING
    4.3.1 Manufacturing and Support Facilities
    4.3.2 Manufacturing Process
    4.3.3 Manufacturer Commentary

4.4 AVAILABILITY OF TRANSPARENCY OF INFORMATION
    4.4.1 Published Resources
    4.4.2 Regular Corporate Sustainability Reports
    4.4.3 Policies on Disclosure
    4.4.4 Manufacturer Commentary

5.0 MANUFACTURER SUPPORT DOCUMENTATION

    5.1 PRODUCT DATA SHEETS
        5.1.1 Website
        5.1.2 Product Listings

GreenFormat™

Structured Format for Information to Support Sustainable Design and Product Choices.

This draft of the "GreenFormat™ version 2.0 Application Guide" is not released for publication; it is only furnished for review to selected participants. Use or distribution of this information for purposes other than commentary on or development of this document is expressly forbidden. This document is not to be considered a final draft or official publication of CSI, and should not be used or cited as such. All contents are copyright © 2014, CSI, 110 South Union Street, Alexandria VA 22314. All rights reserved.

CSI gives acknowledgement and thanks to the members of the GreenFormat Task Team for their work:

George Middleton, CSI, AIA, LEED AP BD+C, GGA (Chair)
Paul Sternberg, CSI, CCS, CCCA, LEED AP, NCARB
Chris Hsieh, LEED AP
Susan Kaplan, CSI, CDT, LEED AP
Lisa Turner Britton, CSI CCPR LEED-AP BD+C
Renee Doktorczyk, CSI, CCS, FAIA, SCIP
Wade Bevier, FCSI, CCS, LEED AP BD+C
Brian Wolf (Tech Com Liaison)
Kirby Davis, CSI, CCTS, CDT, LEED AP BD+C (Board Liaison)

CSI
110 South Union Street, Suite 100
Alexandria, VA 22314
800-689-2900
www.csinet.org

GreenFormat 第一版和第二版（草案）示例

Introduction

CSI's *GreenFormat*™ is a standardized structure for organizing sustainable ('green') information elements associated with materials, products, systems and technologies used in the built environment. By using this standardized format, manufacturers can accurately identify key product characteristics and provide designers, constructors, and building operators with information needed to help meet sustainable design and operation goals. The information collected and organized by using *GreenFormat* can be used to produce a sustainability profile of a product. The identification of the criteria, standards, and applicable certifications by using GreenFormat provides designers, constructors and building operators an easy way to evaluate the various sustainable aspects of materials, products, and processes across a wide variety of manufacturers.

The *GreenFormat* Task Team has organized the structure into specific categories in an effort to make the process of understanding and using sustainability-related product information easy. Information classified according to GreenFormat is grouped into nine categories (with three of those reserved for expansion), each containing individual sub-categories and classifications.

The first three categories allow suppliers and manufacturers to identify information specific to their *product* and these include:

1.0 Product General Information
2.0 Product Details
3.0 Product Lifecycle

Categories 4.0, 5.0 and 6.0 are presently reserved for future expansion.

The last three categories focus on information about the *manufacturer*, which may include environmental policies, environmental or life-cycle assessment initiatives, and social responsibility initiatives. These categories include:

7.0 Manufacturer Sustainability Policies
8.0 Manufacturer Support Documentation
9.0 Manufacturer Certification

The systematic approach of GreenFormat provides manufacturers with a way to communicate the sustainability features of their products and operations, allowing design professionals and other information users to make informed decisions.

This flexible structure is designed to be adaptable to anticipated changes in the industry. As sustainability issues and product selection criteria evolve, new topics may be added in the appropriate category, and existing topics that become obsolete or may change and can be removed as necessary. The structure is designed to support specification writers and the project specification documents they produce by providing the information needed for the four different methods of specifying: Reference Standard, Performance, Descriptive, and Proprietary.

One way to understand the structure and purpose of *GreenFormat* and how it supports the specification writing process is to consider the fact that specifications are typically written with a project requirement stated as a "salient feature" followed by a "value" for that salient feature. For example the salient feature "VOC content" written into a project specification would be accompanied by a value such as "5 grams/liter" which would be one element of information provided by a product manufacturer under a category heading defined under *GreenFormat*.

The organization of *GreenFormat* into specific classifications of information is not intended to imply a hierarchy or relative level of importance of the information within a category. It is the responsibility of the user of the information classified according to *GreenFormat* to determine the suitability of information provided. The user of GreenFormat determines project specific requirements and prioritizes the sustainability criteria by which products will be evaluated. There are many existing evaluation and rating systems available to assist owners, designers, and specifiers in determining the relative contributions of different products to sustainability. The comprehensive structure of *GreenFormat* allows the user to organize the many sources of sustainability information within a hierarchy of both information quality and specificity applicable to each product type. Data providers can organize their data reporting according to GreenFormat categories while retaining their own proprietary features.

One of the key goals of *GreenFormat* is for product information to be transparent and verifiable. A manufacturer is being transparent when the company provides sufficient detailed information about its products and processes so that design professionals can make informed decisions.

GreenFormat emphasizes the importance of objective, science-based and widely-recognized standards and evaluation criteria, arrived at through consensus standard-setting processes. GreenFormat is also designed to work in conjunction with *MasterFormat*™ and can be applied to all construction products and categories.

GreenFormat 第一版和第二版（草案）示例

GreenFormat Structure

GreenFormat has a numbered hierarchical outline structure consisting of Categories and Sub-categories. Following the approach used in MasterFormat, *GreenFormat* is divided into levels, each separated by a period (.). Printed documents may include indents for each level, but it is not required. Information may be listed under different levels of specificity within each category or sub-category. Manufacturers should select the level of specificity appropriate to the information being provided. *GreenFormat* generally includes sub-categories down to level 3 or level 4.

As noted previously, the *GreenFormat* Categories are divided into two groups of three categories each, with the middle three category numbers reserved for future use. Categories 1, 2 and 3 are specific to the product or product category being described. Category 7 is specific to the sustainability efforts of the manufacturer independent of the actual product being described. Categories 8 and 9 are specific to the documentation of the information provided under all other categories.

Use of *GreenFormat* Category Levels

The use of Category levels is intended to be flexible and to provide information detail suitable across a wide range of applications. Each Category is hierarchical, however each Category or sub-category level may, at the option of the information provider, include all information below that level. AGreenFormat product listing is most useful and supports data comparison when information is provided at the appropriate sub-category level. For example, if the product information includes the Volatile Organic Compound (VOC) content, this salient feature may be shown in the following ways:

    Level 1: Performance Criteria
    Level 2: Product Composition (one specific aspect of 'Performance Criteria')
    Level 3: Emissions (one specific aspect of 'Product Composition')
    Level 4: Emission Value (one specific aspect of 'Emissions')

Each subsequent level is intended to be more specific and detailed than the previous level. Use of the appropriate level of information is necessary if multiple sub-categories of information are provided. No category level should contain more than one type of descriptive content. In the example below, listing the VOC content and Urea Formaldehyde content under the same heading would not be correct:

2.3.6 Emissions: Volatile Organic Compound (VOC) Content, 50g/L, no added Urea Formaldehyde

Instead, these properties should be shown as follows:

2.3.6.1 Volatile Organic Compound (VOC) Content: 50g/L
2.3.6.2 Formaldehyde Content: No added Urea Formaldehyde

It is not necessary to include higher level sub-categories as titles when listing information according to *GreenFormat*. If the information is being provided in a readable format, it is recommended that at least the Level 1 titles be used as titles for the sub-category information below it. For example, a minimal listing might be:

2   Performance Criteria:

    2.3.6.1 Volatile Organic Compound (VOC) Content: 50g/L
    2.3.6.2 Formaldehyde Content: No added Urea Formaldehyde

While a complete listing might be:

2   Performance Criteria:

  2.3 Composition of Product:

    2.3.6 Emissions:

    2.3.6.1 Emissions Value: Volatile Organic Compound (VOC) Content: 50g/L
    2.3.6.2 Emissions Value: Formaldehyde Content: No added Urea Formaldehyde

Information that is provided purely as data, such as BIM object data, does not require higher level titles. Blank field entries may be required for data consistency between objects and to provide more readable reports of object attributes.

Category information should be clearly separated from the category title by using a semicolon (:), tab space, or separate cell in a table listing.

Using *GreenFormat* for Data Organization and Comparison

    GreenFormat is not intended to be a prescriptive guide defining how product information should be provided, nor is it intended to be, in and of itself, any particular form of information tool. *GreenFormat* is designed to be an organizational format upon which many potential information tools may be structured. The basis of that structure will almost always be based on elements of information or data.

Manufacturer websites are data driven and designers increasingly seek out websites that allow them to make side-by-side comparisons of product salient features. Sustainability rating systems require input of information into online data bases and evaluation tools. Product cataloging and reporting systems are also dependent on data. While a designer or owner may require printed documents-or electronic facsimiles of those documents-the designers may need access to digital BIM models that allow them to analyze the compliance of their design to project criteria in real time.

GreenFormat addresses these disparate organization needs by providing a structured, numerical outline and hierarchy of information that can be consistently used across all platforms of data delivery. The key component is use of the outline numbering to locate information. It is not necessary to provide a response to every category of information requested in GreenFormat. What is necessary is to use the outline numbering to consistently locate the information that is provided. Those numbers can then be used, for example, to identify specific information elements in a marketing brochure or catalog page. The graphic approach is left entirely up to the manufacturer. Data can be listed in a numbered outline or as fields in a table. Like MasterFormat, the number locates the information within a larger hierarchy.

Electronic approaches to data delivery are even more varied, but even so, the organization data can be consistent with the recommendations provided in *GreenFormat*. Each information element is connected to an outline number and title. The outline numbers then become the key field identifier that can be used to build spreadsheets or databases. Consistent field identifiers are necessary for data sorting and comparisons. The output can be utilized in any form, from stand-alone databases, to spreadsheets, delimited-format data that can be imported into multiple data systems, to web-based applications and information collection or BIM object data. Information in *GreenFormat* categories can be used to populate data under UniFormat Table 49. Design and construction firms that have existing databases of sustainable product information can even map their existing data to *GreenFormat*-based fields that will permit direct importing of *GreenFormat*-based data.

GreenFormat Outline Structure

1.0 PRODUCT-GENERAL INFORMATION

    1.1 MANUFACTURER PROPERTIES

        1.1.1 Company Name

        1.1.2 Subsidiary Name(s)

        1.1.3 Address

        1.1.4 Contact Information

    1.2 PRODUCT PROPERTIES

        1.2.1 *MasterFormat* Number

        1.2.2 Product Identification

        1.2.3 Product Description

        1.2.4 Product Photo(s) or Drawing(s)

2.0 PRODUCT DETAILS

    2.1 Sustainable Standards and Certifications

        2.1.1 Third Party Certification

            2.1.1.1 Whole Product Sustainability

            2.1.1.2 Single Attribute

            2.1.1.3 Other Certification Categories

        2.1.2 Second Party Certification

            2.1.2.1 Whole Product Sustainability

            2.1.2.2 Single Attribute

            2.1.2.3 Other Certification Categories

        2.1.3 Self-Declaration of Compliance

            2.1.3.1 Whole Product Sustainability

            2.1.3.2 Single Attribute

            2.1.3.3 Other Certification Categories

    2.2 PROPERTIES OF SUSTAINABILITY

        2.2.1 Facility Construction

            2.2.1.1 Existing Conditions

            2.2.1.2 Concrete

            2.2.1.3 Masonry

GreenFormat 第一版和第二版（草案）示例

    2.2.1.4 Metals
    2.2.1.5 Wood, Plastics and Composites
    2.2.1.6 Thermal and Moisture Protection
    2.2.1.7 Openings
    2.2.1.8 Finishes
    2.2.1.9 Specialties
    2.2.1.10 Equipment
    2.2.1.11 Furnishings
    2.2.1.12 Special Construction
    2.2.1.13 Conveying Equipment
  2.2.2 Facility Services
    2.2.2.21 Fire Suppression
    2.2.2.22 Plumbing
    2.2.2.23 Heating, Ventilating, and Air-Conditioning (HVAC)
    2.2.2.24 Electrical
    2.2.2.25 Communications
  2.2.3 Site and Infrastructure
    2.2.3.31 Earthwork
    2.2.3.32 Exterior Improvements
    2.2.3.33 Utilities
    2.2.3.34 Transportation
    2.2.3.35 Water and Marine Construction
  2.2.4 Process Equipment
    2.2.4.40 Process Integration
    2.2.4.41 Material Process and Handling Equipment
    2.2.4.42 Process Heating, Cooling, and Drying Equipment
    2.2.4.43 Process Gas and Liquid Handling, Purification, and Storage Equipment
    2.2.4.44 Pollution and Waste Control Equipment
    2.2.4.45 Industry-Specific Manufacturing Equipment
    2.2.4.46 Water and Wastewater Equipment
    2.2.4.47 Electrical Power Generation
  2.2.99 Additional Product Performance
2.3 SUSTAINABLE COMPOSITION OF PRODUCT
  2.3.1 Product Composition
  2.3.2 Chemicals Composition
    2.3.2.1 Unregulated Chemical Content
    2.3.2.2 Regulated Chemical Content

2.3.2.3 Product Content Declarations

2.3.3 Recycled Content

2.3.3.1 Pre-Consumer Material

2.3.3.2 Post-Consumer Material

2.3.4 Rapidly Renewable Materials

2.3.5 Reused Materials

2.3.6 Emissions

2.3.6.1 Volatile Organic Compound (VOC) Content

2.3.6.2 Formaldehyde Content

3.0 PRODUCT LIFE CYCLE

3.1 LIFE CYCLE Assessment

3.2 MATERIAL EXTRACTION AND TRANSPORTATION

3.2.1 Regional Materials

3.2.2 Supply Chain

3.3 MANUFACTURING PROCESS

3.4 CONSTRUCTION

3.4.1 Construction Waste Management

3.4.2 Installation

3.4.3 Contract Closeout

3.5 FACILITY OPERATIONS

3.5.1 Product Service Life

3.5.2 Recommended Cleaning and Maintenance

3.6 REUSE, RECYCLING, DISPOSAL

3.6.1 Manufacturer/Industry Programs

3.6.2 Product Reuse

3.6.3 Product Recycling/Disposal

4.0 RESERVED FOR FUTURE USE

5.0 RESERVED FOR FUTURE USE

6.0 RESERVED FOR FUTURE USE

7.0 MANUFACTURER SUSTAINABILITY POLICIES

7.1 ENVIRONMENTAL STEWARDSHIP

7.2 COPORATE GOVERNANCE

    7.2.1 Employment Policies

    7.2.2 Employer Responsibility

    7.2.3 Community Engagement

    7.2.4 Financial Leadership

7.3 MANUFACTURING

    7.3.1 Manufacturing and Support Facilities

    7.3.2 Manufacturing Process

    7.3.3 Manufacturers Comments

7.9 TRANSPARENCY OF INFORMATION

8.0 MANUFACTURER SUPPORT DOCUMENTATION

    8.1 MARKETING MATERIAL

        8.1.1 Website

        8.1.2 Product Listings

9.0 MANUFACTURER CERTIFICATION

    9.1 AUTHORIZATION OF INFORMATION

        9.1.1 Full Name

        9.1.2 Title

        9.1.3 Authorization

        9.1.4 Date

        9.1.5 Company

            9.1.5.1 Contact Name

            9.1.5.2 Contact Title

            9.1.5.3 Contact Email

            9.1.5.4 Contact Phone

    9.2 TECHNICAL REPRESENTATIVE

        9.2.1 Contact Name

        9.2.2 Contact Title

        9.2.3 Contact Email

        9.2.4 Contact Phone